Foundations of Heterogeneous Integration:
An Industry-Based, 2.5D/3D Pathfinding
and Co-Design Approach

Farhang Yazdani

Foundations
of Heterogeneous Integration:
An Industry-Based, 2.5D/3D
Pathfinding and Co-Design
Approach

 Springer

Farhang Yazdani
BroadPak Corporation
San Jose, CA
USA

ISBN 978-3-030-09323-5   ISBN 978-3-319-75769-8   (eBook)
https://doi.org/10.1007/978-3-319-75769-8

This Springer imprint is published by Springer Nature
The registered company is Springer International Publishing AG
The registered company address is: Gewerbestrasse 11, 6330 Cham, Switzerland

To My Educators

Preface

The semiconductor industry is undergoing a rapid transformation, mainly due to emerging markets and applications. As the CMOS scaling slowdown, industry is relying upon IC packaging industry to help in the development of next generation of low-power and cost-effective products and systems. This trend has created a demand for expert advanced packaging professionals with knowledge and expertise in pathfinding and co-design. Historically, packaging has been considered an afterthought during the product development phase; however, in the past couple of years we have witnessed a rapid transformation in the packaging industry mainly in response to heterogeneous integration demand.

Package design process is transforming from "Through-Over-the-Wall" practice to a tightly coupled co-design process with bulk of the design effort spent upfront to perform pathfinding, feasibility analysis, and product planning. The IC package design practice has evolved over the years, nowadays, in order to perform co-design and interface with the IC design team it is expected that the package design engineer to be familiar and have good knowledge of device floorplan and IC physical implementation. In fact, it is expected that the packaging team to perform early floorplanning investigation and propose an optimal floorplan to ensure optimal package and system connectivity.

The motivation for writing this book was to train and prepare the next generation of engineers and designers needed for the emerging heterogeneous integration era. The book is written based on real-life industry practices with industry-based examples and projects.

This book can be used as a textbook for advanced undergraduate courses or graduate courses or as a supplement to any VLSI course. It can be thought during 1 to 2 quarter or 1 semester as needed. The book is also meant to provide the skills and knowledge necessary for the practicing engineers in various disciplines including packaging engineers who would like to work in the design front.

The book is written in an easy-to-learn format with plenty of graphics and examples to demonstrate the concept. It is assumed that the reader has little or no knowledge of packaging industry. Note that this book focuses on pathfinding, physical implementation, and design aspects of heterogeneous modular integration. This book is organized in the following format:

Chapter 1: Background
Chapter 2: Wirebond design guideline and assembly process
Chapter 3: Flip chip design guideline and assembly process
Chapter 4: Substrate technologies and manufacturing design rules
Chapter 5: Conventional design flow for wirebond and flip chip packages
Chapter 6: Pathfinding and design optimization of 2.5D/3D heterogeneous systems

Author does not endorse any EDA tools to be used to carry out examples and projects in this book. However, reader is encouraged to use any EDA tool available with capabilities to carry out examples and exercises.

Last but not least, I would like to thank Charles Glaser and Brian Halm of Springer for their encouragement and patience during the course of this book.

San Jose, USA   Farhang Yazdani
January 2018

Contents

Chapter 1
Introduction

Abstract In this chapter we will first learn about the background, roadmaps and motivation for advanced packaging and heterogeneous integration followed by introduction to basics of semiconductor packaging.

1.1 Background [1, 2]

The semiconductor packaging industry is driven by innovations in scaling CMOS technology to more advanced process nodes. However, the semiconductor industry has experienced an increase in wafer price, reducing the traditional process shrink benefits beyond 22 nm node. This increase is in part due to increase in equipment cost, extensive process R&D and with lithography contributing to nearly 50% of the overall wafer cost. The slowdown in CMOS scaling and cost constrains has forced semiconductor industry to seek alternative approaches, in response, industry is moving from traditional 2D monolithic System on Chip (SOC) type structure to 2.5D/3D heterogeneous type integration, also known as modular integration (Fig. 1.1). The benefits of 2.5D/3D modular integrations are notably, increase in performance, shorter interconnect length, higher speed, lower delays, reduced power consumption, smaller form factor, reduced weight and volume. More so, 2.5D/3D integration results in reduced costs by allowing integration of mixed node dies from high volume processes and higher reuse of IP cores.

The main thrust for adopting modular integration has been the industry push for IP reuse strategy in order to reduce costs. In IP reuse strategy, the most widely used common embedded IPs are redesigned in the form of a single IP chip also known as IP chiplet/dielet. The IP chiplet in a specific process node has a common interface that can connect and communicate to any other device regardless of the device process node. Figure 1.2 illustrates a monolithic device where major embedded IPs are partitioned into number of fully functional IP chiplets with a common interface; thus, IP becomes independent of core logic. Depending on the IP chiplet bump

F. Yazdani, *Foundations of Heterogeneous Integration: An Industry-Based, 2.5D/3D Pathfinding and Co-Design Approach,*
https://doi.org/10.1007/978-3-319-75769-8_1

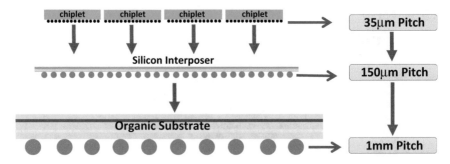

Fig. 1.1 A 2.5D modular integration where an intermediary substrate such as silicon interposer is used to convert finer device bump pitch to coarser package ball pitch. Silicon interpose is used to convert 35 μm bump pitch to 150 μm bump pitch, and organic substrate is used to convert 150 μm bump pitch to 1 mm BGA ball pitch

pitch, a silicon interposer is commonly used as a platform to integrate fine bump pitch chiplets (Fig. 1.1).

Ubiquitous 2.5D/3D heterogeneous integration is evolving as an eminent approach to achieve lower cost, higher bandwidth, smaller footprint, and lower power. However, these benefits come at a price, slicing a 2D monolithic mixed-signal system-on-chip (SOC) die into multiple logic partitions and subsequent heterogeneous integration results in an ultradense off-chip connectivity (Figs. 1.3 and 1.4). Defining, planning, and managing a 2.5D/3D device with ultradense connectivity at the early stages of product development are new challenges that the industry is facing today (Fig. 1.5).

Planning and defining 2.5D/3D devices at the early stages of product development and ultimately architecting a 2.5D/3D device require an advanced pathfinding and optimization methodology, (Fig. 1.6). Such methodology directly impacts the performance, cost, and time to market. Managing and optimizing off-chip ultradense connectivity resulting from 2.5D/3D integration require a novel and unique pathfinding and optimization methodology.

As the 2.5D/3D integration technologies shape the future of semiconductor industry, pathfinding methodology and design optimization of such integration are still at its infancy. Package co-design flow and EDA tool technologies have evolved considerably as a result of such requirement. The modern co-design flow requires an end-to-end system interconnect planning during early stages of product planning. As illustrated in Fig. 1.7, co-design flow includes simultaneous planning and optimization of I/O buffer cell placement, optimization of device bond/bump pad pattern, and optimization of package BGA ball pattern to ensure optimized connectivity with off-the-shelf system components.

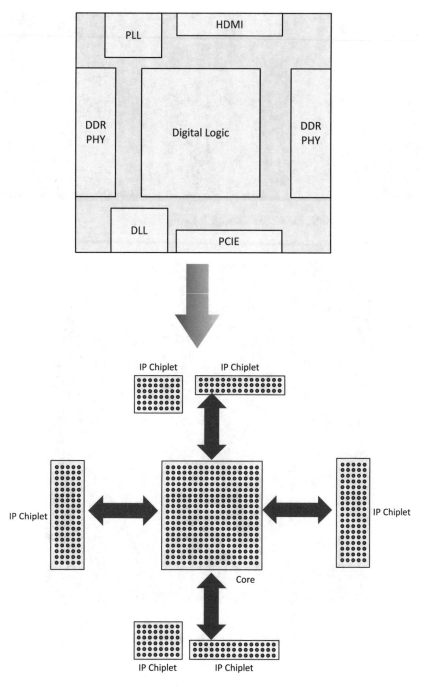

Fig. 1.2 IP reuse strategy where embedded IPs used in a monolithic device are converted into functional chiplets. IP chiplets use common interface to communicate with other devices/cores

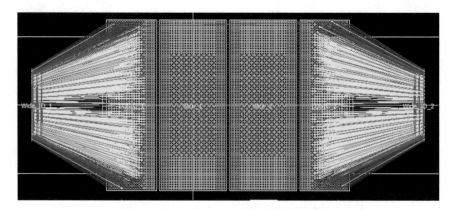

Fig. 1.3 A 2D monolithic die sliced into multiple logic partitions

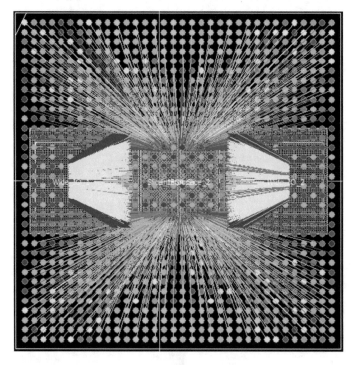

Fig. 1.4 Sliced logic dies in Fig. 1.3 are optimized with respect to silicon interposer and organic substrate BGA ball pattern for optimal connectivity and routing

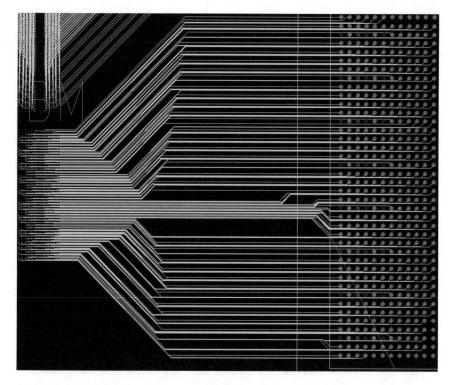

Fig. 1.5 Dense routing connectivity in 2.5D/3D integration

Moreover, modern 2.5D/3D modular integration requires rapid exploration of multiple integration scenarios, multiple devices, multiple physics, and multiple customers (Fig. 1.8).

Connectivity density highly depends on partitioning scheme used to partition the circuit. A device may be partitioned at the I/O level, IP block level, sub-IP level, and cell level as illustrated in Fig. 1.9. Each partitioning scheme results in different pad pitch; for example, partitioning the circuit at the I/O level normally results in bump pitch of 40 μm where die-to-wafer or die-to-interposer packaging technologies are used to integrate the circuit partitions, whereas partitioning a circuit at the standard cell level produces almost 100 nm pad pitch; obviously assembly of circuit partitions with 100 nm pad pitch requires advanced packaging technologies such as active layer bonding process technologies. It should be noted that each of these partitioning schemes results in different cost model; thus, the economics and merit of a partitioning scheme should be investigated in depth and carefully, and the normal rule of thumb dictates that as pitch decreases cost increases.

Emerging markets such as machine learning (ML) have imposed new requirement on packaging industry. Typically, machine learning platform consists of

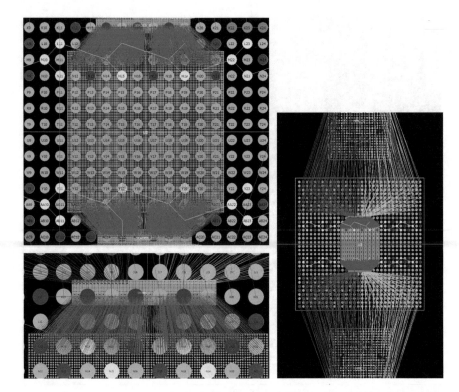

Fig. 1.6 Planning and defining 2.5D/3D devices at the early stages of product development require an advanced pathfinding and optimization methodology. In this figure, parallel interface of the IP chiplet is optimized with respect to the core logic die while its high-speed serial interface is optimized with respect to PCB components with fixed BGA ball pattern

multiple devices integrated with high-end memory device such as high bandwidth memory (HBM) integrated on a silicon interposer. Heterogeneous 2.5D/3D packaging is the key driver for machine learning platforms; in addition, to scale machine learning with Moore's Law heterogeneous 2.5D/3D platform is needed.

The focus of this book is on building the foundation necessary to plan, architect, and design simple to advanced interconnect platforms. Industry-based examples and guidelines are provided to enhance the learning process. We will start with the most basic elements of packaging and then work our way to perform more advanced pathfinding and design optimization. In Chap. 2 we will learn the wire-bond design and assembly process, in Chap. 3 we will learn the flip chip design and assembly process, in Chap. 4 we will learn about substrate technologies and manufacturing design rules and guidelines, in Chap. 5 we will learn about conventional design flow and practices for wirebond and flip chip packages including examples and projects, and finally in Chap. 6 we will learn about defining, planning, pathfinding, and optimization of 2.5D/3D heterogeneous systems.

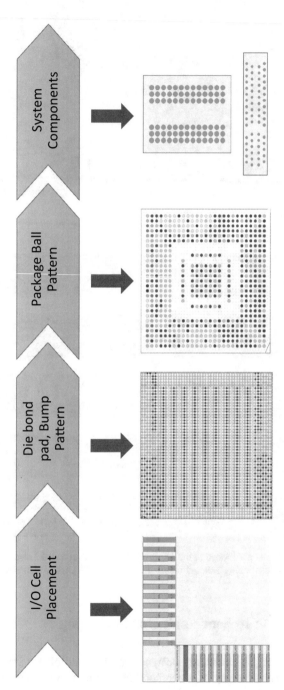

Fig. 1.7 Simultaneous planning and optimization of I/O cell placement, device bond/bump pad pattern, and package BGA ball pattern in the context of off-the-shelf system components

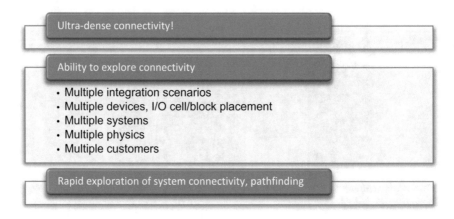

Fig. 1.8 Requirements for modern 2.5D/3D modular integration

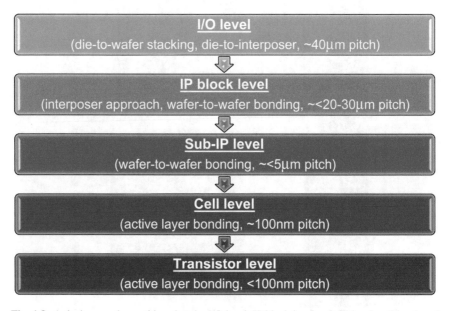

Fig. 1.9 A device may be partitioned at the I/O level, IP block level, sub-IP level, cell level, and transistor level. Each partitioning scheme results in different pad pitch which requires different packaging technologies

1.2   Introduction to Semiconductor Packaging

Complex devices are predominantly designed in wirebond or flip chip configurations (Fig. 1.10). The anatomy of a wirebond and a flip chip package is illustrated in Figs. 1.11 and 1.12, respectively. Wirebond, sometimes referred to as bondwire

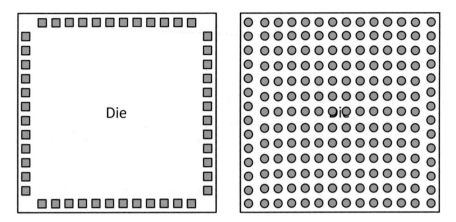

Fig. 1.10 Left, a typical wirebond device where bond pads are placed on the periphery of the device. Right, a typical flip chip device where bump pads are placed on the entire surface of the device and bumped with solder bumps or copper pillars

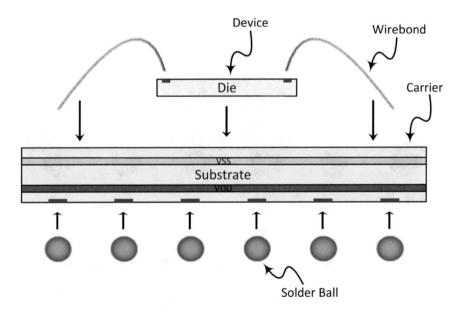

Fig. 1.11 Anatomy of a wirebond package

or simply wire, is the oldest and still widely used method to connect devices, substrates, or components, and it is considered to be more cost effective than flip chip packages. An assembled wirebond and flip chip package are illustrated in Figs. 1.13 and 1.14, respectively. Packages are then assembled on a printed circuit

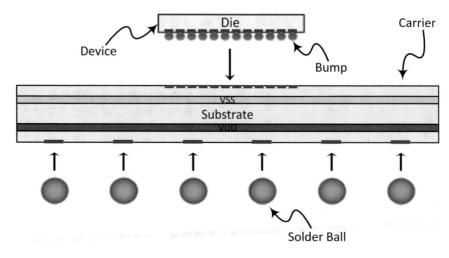

Fig. 1.12 Anatomy of a flip chip package

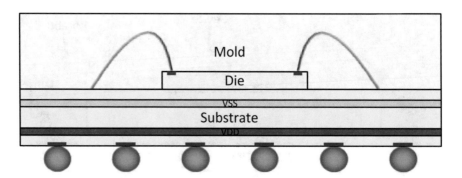

Fig. 1.13 An assembled wirebond package

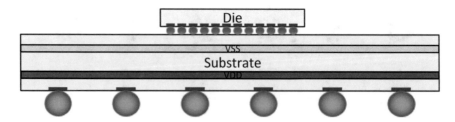

Fig. 1.14 An assembled flip chip package

board (PCB) as illustrated in Fig. 1.15. To understand the structure of typical packages in the industry, we will illustrate the anatomy of a wirebond and a flip chip package.

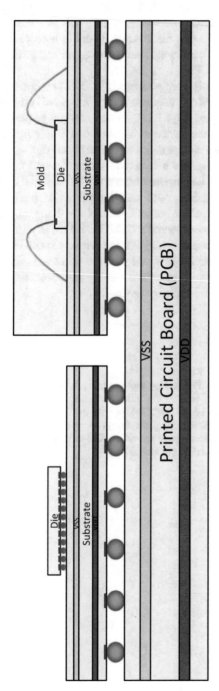

Fig. 1.15 A flip chip and a wirebond package assembled on a printed circuit board

Anatomy of a typical wirebond package (Fig. 1.11) comprises of a device (die), wirebond, substrate (carrier), and ball grid array (BGA) solder balls. During the wirebonding assembly process, die attach material is used to attach the die to the substrate while a molding compound is added to cover and protect the wires and the die as illustrated in Fig. 1.13.

Anatomy of a typical flip chip package (Fig. 1.12) comprises of a device (die), solder bumps or copper pillars, substrate (carrier), and ball grid array (BGA) solder balls. In flip chip packaging (Fig. 1.12), an array of bumps (solder bump or copper pillar) is used to connect the device to the substrate. Flip chip devices are bumped mainly with solder bumps and mounted on the substrate during the assembly, while an under-fill material is disposed between the die and the substrate to fully cover all the bumps with no void. The purpose of the under-fill material is to secure the die to the substrate. In wirebond dies, bond pads are placed on the periphery of the die, while in flip chip dies the entire surface of the die is bumped. Thus, flip chip dies provide more connectivity compared to wirebond dies; in addition, since the back of the flip chip die after assembly (Fig. 1.14) can be exposed to ambient, flip chip dies perform thermally much better compared to wirebond dies.

In the next chapter, we will learn the assembly and manufacturing rules to design substrates for wirebond and flip chip devices.

References

1. Yazdani F, Park J (2014) Pathfinding and design optimization of 2.5D/3D devices in the context of multiple PCBs. In: IMAPS 10th international conference on device packaging, Fountain Hills, AZ, USA, 10–13 Mar 2014, pp 294–297
2. Yazdani F, Park J (2014) Pathfinding methodology for optimal design and integration of 2.5D/3D interconnects. In: Proceedings of the 64th IEEE electronic components and technology conference, Orlando, Florida, 26–30 May 2014

Chapter 2
Wirebond Physical Implementation

Abstract In this chapter, we will learn the basics of wirebonding, including wire selection, design and assembly process. Wirebonding rules and guidelines provided in this chapter are necessary for design of everyday packages as well as advanced packages. Reader is encouraged to master the wirebonding rules and design practices provided in this chapter in its entirety.

2.1 Wirebond Physical Implementation

Wirebond packaging is a mature technology. Wirebonding is a method of connecting a device to outside world using conductive wires. The decision to design a device for wirebond or flip chip packaging is rather involved, and it has more to do with the number of I/Os, signal integrity, power integrity, and thermal integrity requirements with cost consideration in mind. A properly designed and optimized wirebond package can transmit beyond 100 GHz. Wire material can be of gold, copper, silver, or aluminum with wire diameter ranging from 15 to 30 μm. Due to rising cost of the gold, copper wire has become the most widely used lower cost option. For applications where high current-carrying capacity is needed, larger copper wire diameter may be used. The decision to select a certain wire diameter depends on number of factors and is related to the device bond pad opening size as well as bond pad pitch (Fig. 2.1).

Wirebonding can be performed as forward bonding (Fig. 2.2) or reverse bonding (Fig. 2.3) with lower loop profile. However, majority of wirebonding are performed as forward wirebonding.

Before a substrate can be designed, device pad ring must be designed and optimized for the proper package implementation. We will first focus on wirebond type implementation followed by flip chip implementation. As noted earlier, in wirebond devices, bond pads are situated on the periphery of the device. Depending on the bond pad arrangement and signal/supply assignment, appropriate substrate technology and routing strategy must be established. Bond pad arrangement is

© Springer International Publishing AG 2018

F. Yazdani, *Foundations of Heterogeneous Integration: An Industry-Based, 2.5D/3D Pathfinding and Co-Design Approach*,
https://doi.org/10.1007/978-3-319-75769-8_2

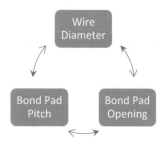

Fig. 2.1 Wire diameter selection is mutually governed by bond pad pitch and bond pad opening size

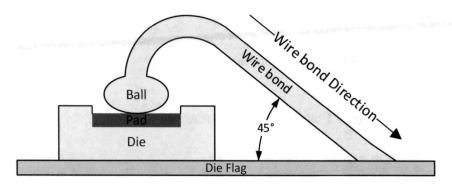

Fig. 2.2 Forward wirebonding

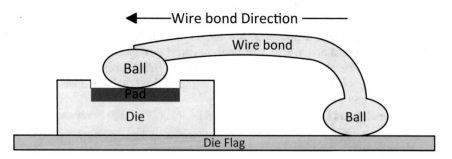

Fig. 2.3 Reverse wirebonding

critical in nature because it must satisfy the device signal and power integrity requirements as well as wirebonding and routing capability of the substrate; thus, establishing an optimal bond pad arrangement for complex wirebond devices can be iterative and ideally requires co-design methodology which we will discuss in more detail in the following chapters.

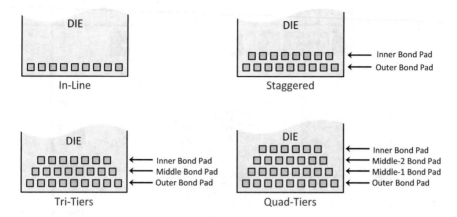

Fig. 2.4 Bond pad placement options for wire bonding process

Most commonly, device bond pad layout is arranged as In-Line or Staggered configuration consisting of outer and inner bond pad rows as illustrated in Fig. 2.4. In some devices, bond pads are arranged in tri-tiers pattern consisting of outer, middle, and inner bond pad rows or quad-tiers pattern consisting of outer, middle-1, middle-2, and inner bond pad rows mainly to support more I/Os and supplies (Fig. 2.4). As a normal rule for wirebond assembly, bond pads on the outer rows are assigned to supplies (power, ground) and inner rows are assigned to signals.

In order to design a device for wirebonding assembly process, certain design rules must be followed. Design rules for in-line bond pad layout is illustrated in Fig. 2.5 and Table 2.1. All parameters for bond placement are tabulated in Table 2.1; it is important to notice that these values are the minimum values. Depending on the die size and number of I/Os and supplies, bond pads of various pitch can be selected from Table 2.1. Bond pads can be of various forms and sizes, but most commonly used are square or rectangular shape. It is worth noticing the difference between the corner bond pad pitch compared to center bond pad pitch. Coarser bond pad pitch at the corners is required to prevent wire crossing during the assembly process and is essential to the design. Typically, minimum of four coarser bond pads on the corners next to clearance zones are required according to Fig. 2.5. Similarly, design rules for staggered bond pad layout are illustrated in Fig. 2.6 and Table 2.2. All parameters for bond placement are tabulated in Table 2.2.

It is the package design team responsibility to communicate interactively with the IC design team and to make sure that the device final bond pad placement as designed by the IC design team is in compliance with the design rules as illustrated in Figs. 2.5 and 2.6 as well as Tables 2.1 and 2.2. For practical purposes, the most recent bond pad placement design rules must be obtained from the assembly subcontractor. Likewise, bond pad layout guideline for higher tier wirebond design can be obtained from the assembly subcontractor.

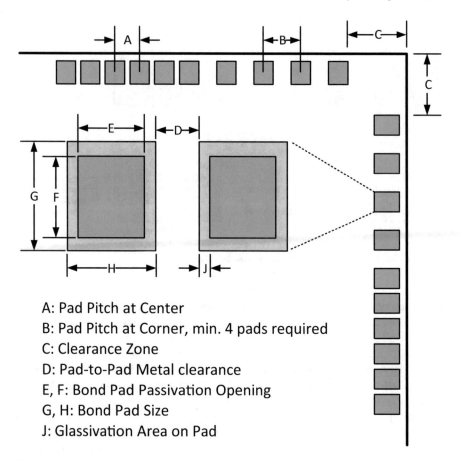

A: Pad Pitch at Center
B: Pad Pitch at Corner, min. 4 pads required
C: Clearance Zone
D: Pad-to-Pad Metal clearance
E, F: Bond Pad Passivation Opening
G, H: Bond Pad Size
J: Glassivation Area on Pad

Fig. 2.5 Typical in-line bond pad layout rules

Table 2.1 Suggested in-line bond pad layout specifications in accordance with Fig. 2.5

A	70	60	55	50	45	40
B	82	82	82	80	80	80
C	150–240	150–240	150–240	150–240	150–240	150–240
D	5	5	4	4	3	3
E	59	50	48	43	38	35
F	63	56	55	55	55	55
G	61	60	60	60	60	60
H	60	54	50	44	41	37
J	3	3	2	2	2	1
Suggested max. wire diameter	30	25	23	20	20	15

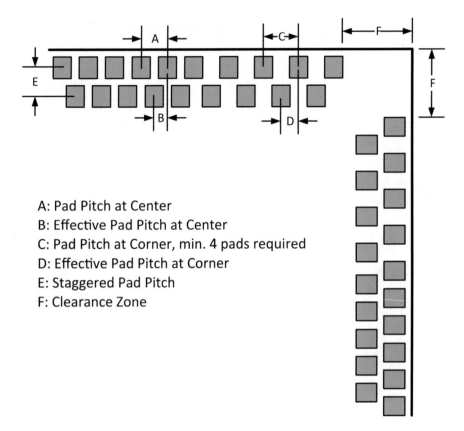

A: Pad Pitch at Center
B: Effective Pad Pitch at Center
C: Pad Pitch at Corner, min. 4 pads required
D: Effective Pad Pitch at Corner
E: Staggered Pad Pitch
F: Clearance Zone

Fig. 2.6 Typical staggered bond pad placement design rules

Table 2.2 Suggested staggered bond pad layout specifications in accordance with Fig. 2.6	A	110	100	90	80	60	50
	B	55	50	45	40	30	25
	C	120	120	110	110	110	90
	D	60	60	55	55	55	45
	E	105	95	85	85	75	75
	F	240	240	240	240	240	240

2.1.1 Wirebond Design

Designing a wirebond substrate requires detailed attention to numerous design rules and specifications. More importantly, wirebond diameter and length play an important role on device performance such as signal integrity and power integrity. For critical memory interfaces as well as high-speed serial interfaces, wirebond

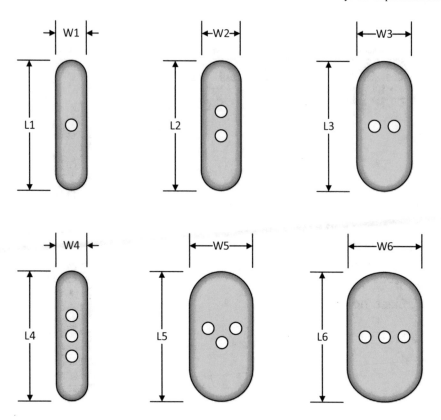

Fig. 2.7 Various bond-finger dimensions to accommodate one or more wires based on design intent. All units are in micron

length must be well controlled and minimized. Depending on the bond pad pitch and bond pad opening size, proper wire diameter must be selected. For staggered and multi-tier bond pad layout, wire loop height control plays an important role on the wirebond ability of the device during the assembly process. Each wire emanates from the device bond pad and lands on a bondfinger on the substrate. Bondfinger come in various sizes and forms, but the most common type as practiced in the industry is the oblong shape illustrated in Fig. 2.7. Bondfinger width and length vary in order to provide sufficient landing surface for one, two, three, or more wires attached to the same bondfinger. For example, to accommodate a single wire, bondfinger can have minimum of 100 μm in width and 260 μm in length. To land two wires vertically or horizontally with respect to each other, bondfingers with the width/length of 110/300 and 210/260 μm can be used, respectively, according to Fig. 2.6. Bondfinger pitch varies and it depends on the manufacturer's capabilities; however, bondfinger pitch of 140 μm for single wire attachment is within reach of most substrate manufacturers, and more aggressive pitch of 120 μm or less should be discussed with the manufacturer. Bond finger dimensions provided in Fig. 2.6 can be used as a guideline to design appropriate bond fingers.

Bondfingers can be patterned with various styles on the substrate; however, the most common types are arc and in-line styles as illustrated in Fig. 2.7. Bondfingers are typically solder mask defined; solder mask coverage of bondfinger according to Fig. 2.7 is about 50 μm on each end of the fingers.

2.1.2 Wirebond Layout Rules

Wirebonding a device to a substrate requires placement of the device on the substrate followed by sequential placement of one or more rings on the periphery of the device on the substrate. Normally, the ring closest to the device is assigned as ground ring, followed by power domain rings such as core and I/O powers. A ring can be partitioned into number of smaller rings in order to support multiple power/ground domains; in addition, bondfingers can be situated on the ring as well.

As noted earlier, bondfingers can be placed at various locations on the substrate as long as it does not violate the minimum and maximum wire length and bondfinger-to-bondfinger spacing or spacing violation with respect to other metals and components. Typically, wire length varies from minimum of 0.5 mm to maximum of 7.5 mm. In addition, there are number of other bonding rules that the designer must be vigilant about. In order to bond all the device bond pad openings and to avoid maximum wire length violation and minimum spacing violations, it is often desired to accommodate bondfingers on two shelves as depicted in Fig. 2.8. As a result, one of the most common design violation scenarios encountered in the design process is wire crossing adjacent bond pad opening especially at the corners. Shorter wire may not cross adjacent corner of the bond pad opening of a longer wire, and this scenario is illustrated with a red forbidden sign in Fig. 2.8. On the other hand, longer wire may cross adjacent corner of the bond pad opening of a shorter wire given that at least one wire diameter clearance (W1) exists between the two wires, and this scenario is illustrated with a green checkmark in Fig. 2.8.

In multi-shelves bondfinger configuration, clearance of at least 1 wire diameter is required between the inner row bondfinger and adjacent wires as illustrated in Fig. 2.9. Similarly, wire crossing adjacent bondfinger corner should be avoided, and care should be taken to make sure that wire direction and bond fingers are as parallel as possible as illustrated in Fig. 2.10.

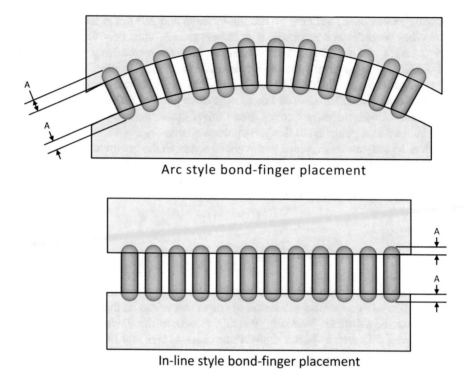

Arc style bond-finger placement

In-line style bond-finger placement

Fig. 2.8 Bond-finger placement using arc and in-line style with solder mask coverage of $A = 50$ μm

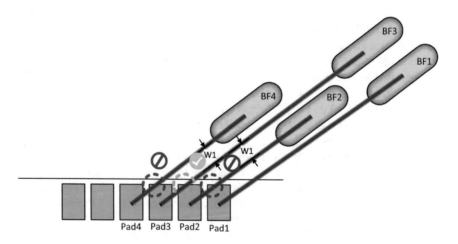

Fig. 2.9 Wire crossing the adjacent bond pad opening is forbidden. That is, wire emanating from Pad4 may not overlap the corner of Pad3. Exception: In a situation where wire length connected to bond finger BF3 is greater than or equal to wire length connected to bond finger BF2, then wire connected to BF3 may cross the adjacent bond pad opening (Pad2) if at least one wire diameter clearance (W1) exists between the two wires

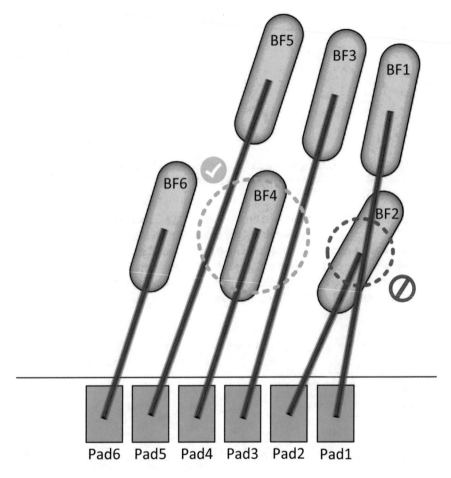

Fig. 2.10 Wires and bond fingers on multi-shelves arrangement must be parallel and aligned. Wire connected to bond-finger BF1 may not cross wire/finger BF2. Clearance of at least 1 wire diameter is needed between the inner row bond-finger and adjacent wires. That is, edge-to-edge spacing between bond-finger BF4 and adjacent wires must be at least 1 wire diameter

2.1.3   Wire Loop Control

Distance from die bond pad surface to the wire's highest point is referred to as wire loop height. Wire loop profile is one of the critical design parameters that must be observed during wirebond design process. In order to avoid wire short, different wire loop heights are needed. Figure 2.11 illustrates wirebonding a die with in-line bond pad configuration to a substrate, and design parameters such as wire diameter and bond pad pitch are indicated on each wire group. Device ground pads (vss) are bonded to vss ring-1 with wire loop height of 7 mil, power pads (vdd1) are bonded to vdd-1 ring-2, power pads (vdd2) are bonded to vdd-2 ring-3 with wire loop height of

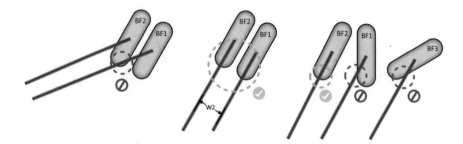

Fig. 2.11 (Left) Wire crossing adjacent bond-finger corner is forbidden. (Middle, Right) Wire direction and bond-finger must be as parallel as possible

12 mil, while signals are bonded to finger-1 and finger-2 shelves with loop heights of 15 and 17 mil, respectively. Similarly, for a device with staggered bond pad layout, wire loop height parameters and bonding patterns are illustrated in Fig. 2.12. Note that die inner row bond pads are connected to outer row bondfingers on the substrate. Wire loop parameters and bonding profiles for tri-tiers and quad-tiers bond pad layout are illustrated in Figs. 2.13, 2.14 and Table 2.3, respectively. As a normal

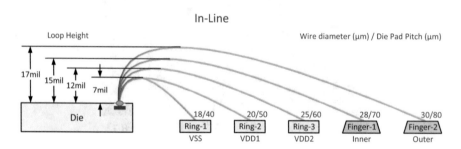

Fig. 2.12 Wire loop height, wire diameter, and bond pad pitch guidelines for wire bonding a device with single bond pad row to substrate rings and fingers

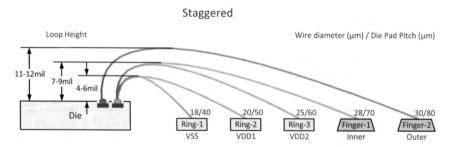

Fig. 2.13 Wire loop height, wire diameter, and bond pad pitch guidelines for wire bonding a device with staggered bond pad rows to substrate rings and fingers

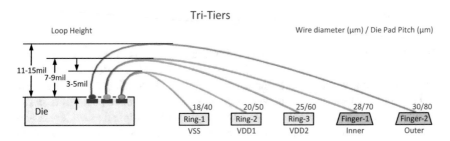

Fig. 2.14 Wire loop height, wire diameter, and bond pad pitch guidelines for wire bonding a device with tri-tiers bond pad rows to substrate rings and fingers

Table 2.3 Wire loop height guidelines for wire bonding a device with quad-tiers bond pad rows to substrate rings and fingers. Die thickness is assumed to be 14 mil

Loop height (unit: mil)	Outer pad	Middle pad 1	Middle pad 2	Inner pad
Ring-1 (VSS)	3–4	5–6		
Ring-2 (VDD1)		6–8	9–10	
Ring-3 (VDD2)		6–8	9–10	
Finger-1 (inner)			9–10	12–13
Finger-2 (outer)			11–12	15–16

design practice, wire clearance among different bond pad rows is controlled to at least one or two wire diameter. Most modern EDA tools are equipped with wirebond profile and loop control capabilities as well as 3D design rule check (DRC) verification capabilities to ensure that wirebond design is in compliance with the assembly requirements.

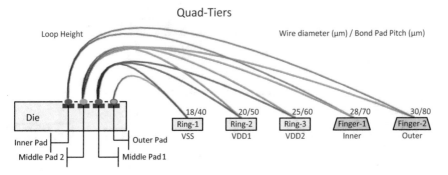

Fig. 2.15 Wire loop height, wire diameter, and bond pad pitch guidelines for wire bonding a device with quad-tiers bond pad rows to rings and fingers. Wire clearance between different pad row wires should be at least 1 wire diameter. Die thickness of 14 mil is assumed

2.1.4   Power and Ground Ring Structure

Normally, device power and ground pad openings are wirebonded to the substrate power and ground rings. Figure 2.15 illustrates top view of a device with in-line bond pad layout wirebonded to the substrate rings and bondfingers, and green and red colors depict ground and power supplies, respectively. Specifications to design substrate rings are illustrated in Fig. 2.16. Rings can be designed as solder mask or non-solder mask defined (Fig. 2.17). Normally, die flag (metal pad under the die) is

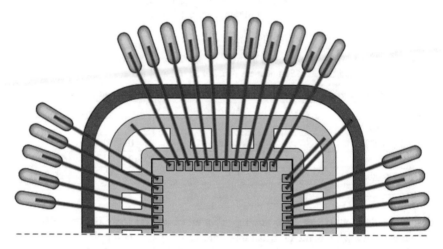

Fig. 2.16 An example of wire bond layout (top view), depicting bond-fingers placement, power ring (red), and die flag combined with ground ring (green). Device power and ground pad openings are wire bonded to the substrate power and ground rings, whereas signals are wire bonded to bond-fingers

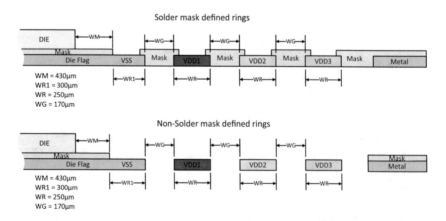

Fig. 2.17 Cross-sectional view, depicting solder mask defined and non-solder mask defined ring structures

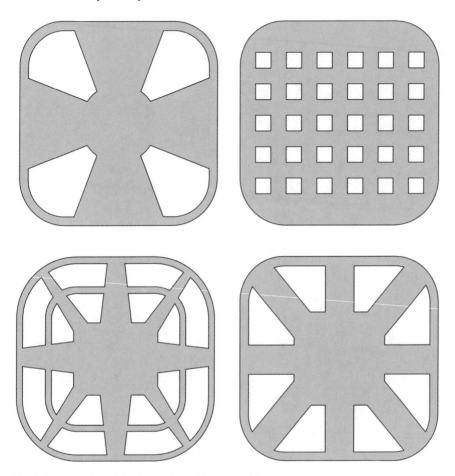

Fig. 2.18 Examples of die flag and ground ring combined

combined with the ground ring, simply called die flag. Die flags can be designed with various structures as illustrated in Fig. 2.18 to provide thermal stress relief for the device.

Chapter 3
Flip Chip Physical Implementation

Abstract In this chapter, we will learn the basics of flip chip packaging, including bump pattern selection, design, and assembly process. Flip chip rules and guidelines provided in this chapter are necessary for design of common flip chip packages as well as advanced packages. Reader is encouraged to master the flip chip design strategy and rules provided in this chapter.

3.1 Flip Chip Physical Implementation

Flip chip packaging is a mature technology. Flip chip, also known as controlled collapse chip connection (C4), is a method of connecting a device to outside world using conductive bumps. Bumps are mostly made up of leaded solder, lead-free solder, or copper materials with various sizes and shapes. Copper bumps are also known as copper pillars or copper posts which are used to bump fine pitch devices. Normally, devices with bump pitch of less than 130 μm are bumped with copper pillars. Copper pillars are mostly constructed with circular cross section; however, oblong shape cross section is also used to increase routing density.

Flip chip devices are typically assembled on a multi-layer high-density substrate. Depending on the bump pitch and bump pattern, appropriate substrate technology, fan-out, and routing strategy must be established. In this section, we will explore bump pattern structure and fan-out strategy used to design common substrates for flip chip devices. Bump pattern is critical in nature because it must satisfy the device signal and power integrity requirements as well as fan-out routing capability of the substrate; thus establishing an optimal bump pattern to satisfy the system performance as well as cost requirement is an iterative process and ideally requires co-design methodology which we will discuss in more detail in the next chapters. From cost perspective, the goal is to escape route and design the substrate with minimum number of layers while maintaining signals and power integrity requirements.

© Springer International Publishing AG 2018
F. Yazdani, *Foundations of Heterogeneous Integration: An Industry-Based,
2.5D/3D Pathfinding and Co-Design Approach*,
https://doi.org/10.1007/978-3-319-75769-8_3

3.1.1   Under the Die Escape Strategies

One of the major tasks in flip chip design is to plan and investigate fan-out and escape routing strategies under the die. The ability to escape route highly depends on the device bump pattern and pitch as well as substrate line width/spacing; finer line width/spacing translates to ability of routing more signals in between the bumps. In a scenario where signal bumps are located on the outer rows of the device, signals may be escape routed on the substrate top layer as depicted in Fig. 3.1. Figure 3.1 shows portion of a substrate where outer row bumps are routed straight off the device, and next second row bumps are routed in between the outer row bumps. This means the two outer row bumps can be escaped on the substrate top layer. Thus, maximum routing density is achieved by patterning the signal bumps on the outer two rows of device. Obviously, with finer trace width/spacing a third row of signal bumps may be escaped through the outer two rows and routed on the substrate top layer. However, it must be noted that finer trace width/spacing increases the cost substantially and signal/power integrity may be effected as well. Thus, planning for such design requires substrate manufacturer's fabrication capabilities, cost consideration, and signal/power integrity simulation and analysis.

Once the routing strategy on the first layer of the substrate is established, escape for the remaining power, ground, and signal bumps must be investigated and strategized for routing on the subsequent substrate layers. Power and ground bumps are dropped using vias to substrate internal power and ground planes, normally, one

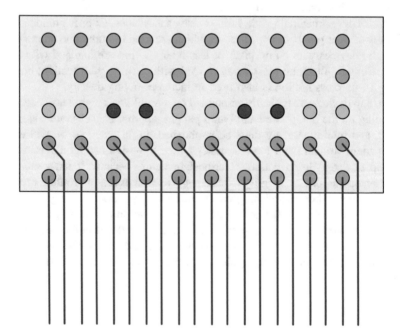

Fig. 3.1 Under the die, escape routing of outer row bumps. Outer two rows signal bumps are routed on the substrate top layer

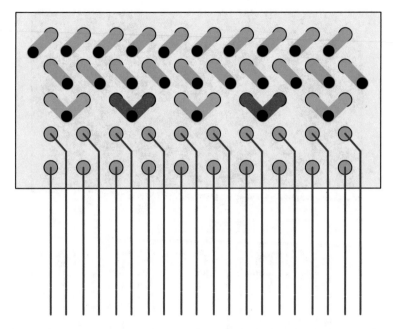

Fig. 3.2 Power and ground bump pads are dropped to substrate internal layers using vias, typically one via for every two bumps. Inner signal bump pads are dropped from layer 1 to layer 2 using vias

via for every two bumps is adequate as depicted in Fig. 3.2. Vias may be placed on the substrate bump pads, widely known as via-on-pad style or offset from bump pad widely known as dog-bone style. Vias can be used to drop next set of signal rows (Fig. 3.2) to second substrate layer and route as illustrated in Fig. 3.3.

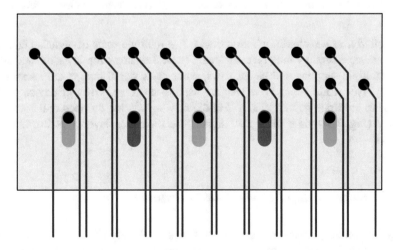

Fig. 3.3 Inner rows signal bumps depicted in Fig. 3.1 are dropped from layer 1 to layer 2 and routed on layer 2

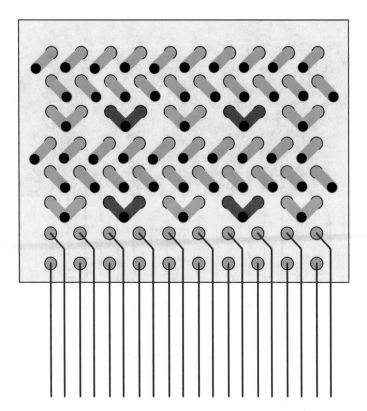

Fig. 3.4 Portion of a device with six signal bump rows and power/ground bump rows in between. Depicting, under the die escape routing of outer row signal bumps. Outer two rows of signal bumps are routed on the substrate top layer. Power and ground bump pads are dropped to substrate internal layers using vias, typically one via for every two bumps. The middle signal bump pads are dropped from layer 1 to layer 2 and inner bump pad signals are dropped to layer 3 for routing

Similarly, in a scenario, where a device has multiple rows of signals (Fig. 3.4) similar escape strategy may be exercised. Figure 3.4 shows top layer of a substrate with multiple supplies and signal rows, outer rows signal bumps are escaped on the top layer while appropriate vias are used to drop next rows of device supply bumps to substrate supply planes. Middle row signals are dropped and routed on layer 2 (Fig. 3.5) while inner row signals are routed on layer 3 as illustrated in Fig. 3.6.

3.1.2   Other Bump Pattern Considerations

As noted, bump patterns and escape strategies are highly driven by signal/power integrity requirements as well as substrate manufacturing capabilities. There are

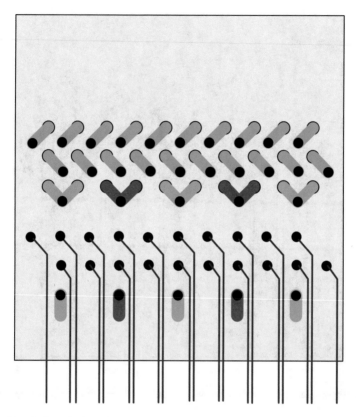

Fig. 3.5 Signals from device middle rows are dropped to substrate second layer and routed

many alternative bump patterns that are practiced in the industry, many driven by specific IP electrical requirements, in particular, low voltage differential signals (LVDSs) as well as 50 Ω signals. Differential signals can be routed as stripline or micro-strip on any layer, however, to achieve maximum routing density, differential bump pairs are assigned perpendicular to die edge, this allows side-by-side escape routing during fan-out. Routing 50 Ω signals pose a bit of challenge, especially in a design that is predominantly 50 Ω, such design to be routed as stripline requires power/ground bumps to be patterned on the edge of the device (Fig. 3.7). This allows implementing a solid voltage reference plane on the top layer of the substrate.

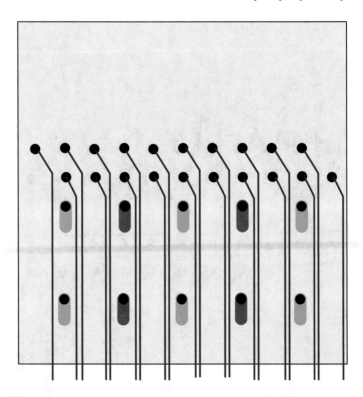

Fig. 3.6 Signals from device inner rows are dropped to substrate third layer and routed

Fig. 3.7 A common 50 Ω
LVDS bump pattern with
ground bumps on the
periphery of the device

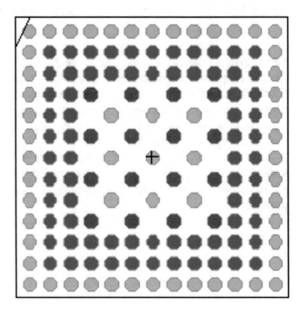

3.1.3 Core Power and Ground Bump Pattern

Having described a common bump pattern where signal bumps are assigned on the periphery of the die for ease of routing escape on the top layer of the substrate, we now focus on the core power and ground bump pattern of the device. One of the key criteria in a successful high-yield flip chip assembly is to ensure an even array of bump distribution across the device, this is important for two chief reasons, first it helps to provide an even flow of underfill during the assembly process, second it helps to evenly distribute the stress during assembly and reduce warpage. There are two common types of core bump pattern, first pattern is a uniform full array with the similar bump pitch as the rest of the bumps on the die, second pattern is a 50% depopulation array as illustrated in Fig. 3.8. As observed so far, the common practice is to allocate signal bumps around the periphery of die while the core is dedicated to power and ground bumps.

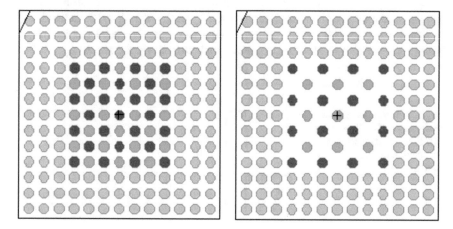

Fig. 3.8 Uniform pitch, full array core bump pattern (left), 50% depopulated core bump pattern (right)

3.1.4   Device Orientation and Fiducials

Plans for placement and orientation of the device is discussed at the early stages of the device physical implementation just to make sure that necessary features are in place so that the device can be assembled on the substrate. From the assembly perspective, device placement machines use optical system to verify the device bump alignment and placement. In order to verify the die alignment prior to placement, the vision system looks for a non-symmetric bump pattern on the device. The common practice is to remove a bump from any location on the device except from exact center of the die; preferred missing bump is typically die pin 1 corner.

In order to align the die on the substrate, certain fiducial features must be designed on the top layer of the substrate. Figure 3.9 shows typical fiducials used in flip chip substrate design. Normally, four fiducials should be present on the four corner of die (Fig. 3.10). One fiducial must be unique and different than other three to identify the package A1 ball corner; normally, a cross shape is used. Fiducials should be placed with center of each fiducial at a distance of 0.2 mm from die edge in any direction. Furthermore, fiducials are typically solder mask defined with typical overlap of 100 μm over a large metal pad, with the metal pad having the same surface finish as all other exposed metal surfaces on the substrate.

Fig. 3.9 Common fiducials designed on the substrate top layer for optical placement recognition during assembly process

Fig. 3.10 Typical fiducial placement around the device with cross shape to identify the package A1 ball corner

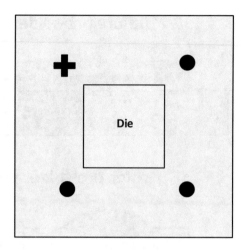

Table 3.1 Typical flip chip pad design guideline

Bump pitch	Pad size	Solder mask opening
260	150	100
230	140	90
200	130	90
180	125	80
140	110	75

3.1.5 Flip Chip Pad Design

Flip chip pad size selection depends on the bump pitch and should be solder mask defined. Typical bump pitch, pad size, and mask opening are tabulated in Table 3.1.

3.1.6 Flip Chip Surface Finish

Flip chip pads need a proper surface finish treatment and preparation before flip chip assembly process. There are number of surface finish options available each with different cost and reliability model, just to name a few popular options, these include:

- Electroless Nickel Immersion Gold (ENIG),
- Electroless Nickel Electroless Palladium Immersion Gold (ENEPIG),
- Immersion Tin (IT),
- Organic Solderability Preservatives (OSP).

In addition, solder must also be applied to the flip chip pad surface, widely known as solder on pad (SOP) (Fig. 3.11). After pre-solder process, top of the solder must be flattened, widely known as coining process as illustrated in Fig. 3.11, coined flat solder diameter typically ranges between 50 and 140 μm. Normally, solder mask above the flip chip pad should have a thickness of 20 μm (Fig. 3.11). Top of the

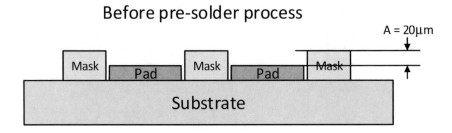

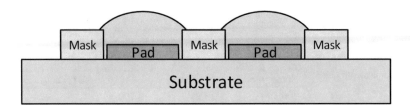

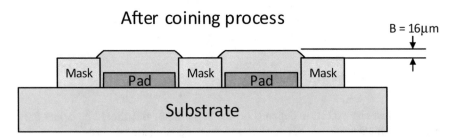

Fig. 3.11 Solder on pad (SOP) coining process as part of substrate manufacturing. Typically, solder mask thickness of $A = 20$ μm above the flip chip pad prior to pre-solder and solder thickness of $B = 16$ μm above the solder mask after coining process is desired

solder after coining process must be at least 16 μm above the solder mask as depicted in Fig. 3.11, and the overall coplanarity of coined surfaces must be less than 25 μm (Fig. 3.11).

3.1.7 Substrate BGA Pad Design and Orientation

Depending on the BGA ball pitch, proper landing pad, and mask opening size are required. Typical parameters to select proper pad size and solder mask opening size based on specific ball pitch are tabulated in Table 3.2. Other variations of pad size

Table 3.2 Typical BGA land pad design and mask opening guideline

BGA ball pitch	Pad size	Solder mask opening
1270	670	540
1000	630	500
800	450	350

Fig. 3.12 Orientation fiducial normally in the form of a triangle must be present on the BGA side of the substrate to identify the package A1 ball corner. A1 ball fiducial is normally solder mask defined over a larger metal and should have a clearance of 500 μm from other metals at the vicinity to prevent possible short

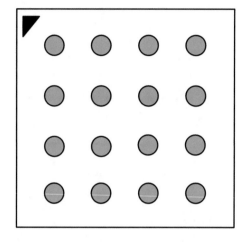

and mask opening for a specific ball pitch are possible pending approval from assembly subcontractor. Similar to the flip chip die side fiducial requirement as discussed earlier, BGA side of the substrate also requires a fiducial to identify the package A1 ball corner and orientation as illustrated in Fig. 3.12. A1 ball fiducial is typically in the form of a triangle with at least 500 μm clearance to any metal feature at the vicinity to prevent possible short. Fiducial is typically solder mask defined over a larger triangle metal shape.

Chapter 4
Substrate Physical Implementation

Abstract Substrate is the foundation of semiconductor packaging. In this chapter, we will learn about various substrate technologies, structures, and manufacturing design rules. We will also learn about substrate stack-up set-up routing consideration. We will discuss silicon interposer manufacturing process and its application including interposer size, TSV aspect ratios.

4.1 Substrate

Substrate is the foundation of semiconductor packaging. Substrate provides the mechanical support as well as an electrical medium to connect one or more devices. Substrate role is to fan-out the device fine pad pitch to a coarser pad pitch and most importantly provide a pathway for the device to communicate with the rest of the system. Substrate is critical in nature, because it has the first-order effect on device performance as well as total package (product) cost. More than 40% of the package cost is attributed to the substrate cost. Substrates are custom designed based on a unique device bond pad or bump pattern and in many cases based on predefined BGA ball pattern. Substrate design cycle spans from weeks to months followed by a lengthy manufacturing process that on average takes between 4 and 8 weeks. In addition, if substrates are not designed optimally one or more re-spin will be required with the similar lead-time as the original build.

Substrate selection and design are one of the most critical aspects of packaging. Substrate type and cross section (stack-up) highly depend on:

1. Bump pitch
2. Routing density
3. Electrical requirement (IP interface)
4. Overall thickness

Depending on the device requirements, substrates can be architected and designed for wirebond or flip chip dies. Mainstream and the most popular substrates are ceramic and organic types, advanced substrates to support finer bump pitch can be

F. Yazdani, *Foundations of Heterogeneous Integration: An Industry-Based, 2.5D/3D Pathfinding and Co-Design Approach,*
https://doi.org/10.1007/978-3-319-75769-8_4

Fig. 4.1 Organic substrate categories

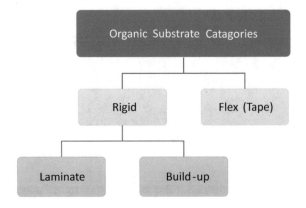

made of glass, silicon, various silicon compounds, thin films, etc. Vast majority of substrates in the market are organic based, supporting device bump pitch of greater than 150 μm and trace width/spacing of greater than 20 μm, whereas silicon substrate (silicon interposer) is widely used to fan-out devices with finer bump pitch, typically less than 100 μm and finer trace width/spacing, typically less than 10 μm.

Among the variety of substrates, major categories are organic based rigid substrates and flex substrates also known as tape substrates (Fig. 4.1). Substrates are classified in terms of their mechanical strength. Rigid substrates are mechanically rigid and thicker, while flex substrates are flexible and thinner. Flex substrates are made of high strength and temperature polymeric materials such as polyimide. Rigid substrates can be made of ceramic, organic, or other materials, but our focus is on organic rigid type substrates. Organic rigid substrates are classified into two categories, laminate and build-up types.

In this chapter, we will study some of the most common substrates used in the industry for wirebond and flip chip packaging, such as laminate, build-up and strategies for stack-up and routing consideration. We will also study the more advanced silicon substrate also known as silicon interposer used to fan-out very fine pitch bump patterns.

4.1.1 Laminate Substrate

Laminate substrates are considered a low-cost substrate in the industry, widely used in wirebonded BGA packages to produce low-cost packages. Laminate substrates are constructed by stacking layers of thin laminate materials with double-sided copper foil. Common laminate materials are FR4 which is epoxy-based and bismaleimide-triazine (BT) which is a higher performance more advanced resin-based material. For packaging applications, BT material is the preferred option and has become the standard material for BGA substrates.

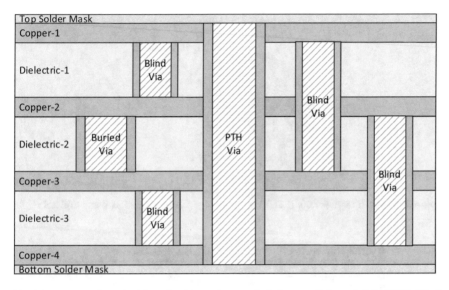

Fig. 4.2 Cross section of a 4-layer laminate substrate, depicting punch through hole (PTH), blind and buried vias. Laminate substrates are widely used for low-cost wirebond packaging

A laminate substrate can have several layers of stacked laminates such as 2, 4, 6 layers, or more. Fig. 4.2 illustrates cross section of a typical 4-layer laminate substrate. Connection among the layers in vertical direction is achieved using vias. Vias of various type can be used to connect the layers and to route and transition signals or supplies among the layers. Vias can be broadly defined as blind, buried, and through holes, Fig. 4.2. Blind via connects the outer most layer to any internal layer, for example, in a 4-layer cross section a blind via can start form top layer (copper-1) and terminat on either internal layers such as copper-2 or copper-3 layers. On the contrast, buried via only connects the internal layers, hence internally buried within the substrate, for example as illustrated in Fig. 4.2 buried via connects copper-2 and copper-3 layers. Through hole, plated through hole (PTH), or punch through hole (PTH) all means the same, as the name suggest, this type of via runs all the way through the substrate from top layer to bottom layer. PTH via consists of a drill hole and a capture pad also known as via land pad as depicted in Fig. 4.3. Common dimensions for PTH pad/drill sizes are 400/200 and 300/150, and these parameters highly depend on the thickness of substrate.

A substrate design consisting of only PTH via is highly desired and considered to be the lowest cost substrate, factors that further effects the costs are PTH via diameter and drill hole size, details of these dimensions and percent increase in costs should be worked out with the substrate manufacturer. As the substrate features get smaller cost increases, this is true for any feature of the substrate such as trace width and spacing. (Figs. 4.4 and 4.5).

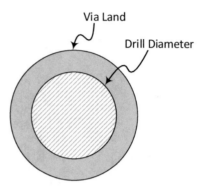

Fig. 4.3 A punch through hole (PTH) via, consisting of land (capture pad) and drill hole

Overall Thickness Options: 170, 210, 260, 360, 560μm

Top Solder Mask: 25, 30μm		
Copper-1: 12, 15, 18, 22μm		
Dielectric-1: 70, 100, 150, 250, 400μm	PTH Via	
Copper-4: 12, 15, 18, 22μm		
Bottom Solder Mask: 25, 30μm		

Fig. 4.4 Cross section of a 2-layer laminate substrate. Thickness of each layer is indicated within the layers, depending on the layer thickness selection, various total substrate thicknesses such as 170, 210, 260, 360, and 560 μm can be achieved

4.1.2 Build-up Substrate

Build-up substrates are constructed by sequentially adding thin non-reinforced epoxy layers (build-up layers) on both sides of a thick laminate core, Fig. 4.6. A thick core is used for mechanical support and as a base to add the build-up layers on both sides symmetrically. A build-up substrate is defined by a core and number of build-up layers. A 4-layer build-up substrate is denoted as 1 + 2 + 1 or 1/2/1, where 1 means one build-up layer and 2 means a laminate core with 2 metal layers, one metal layer on the top and one metal layer on the bottom of the laminate core, Fig. 4.6. Similarly, a 6-layer build-up substrate is denoted as 2 + 2 + 2 or 2/2/2 meaning 2 build-up layers on each side of a 2-layer laminate core, Fig. 4.7 and a 3 + 2 + 3 means 3 build-up layers on the top and 3 build-up layers on the bottom of a 2-layer laminate core resulting in 8-layer build-up substrate. For example, 4 + 2 + 4 and 5 + 2 + 5, 6 + 2 + 6 translate to 10, 12 and 14-layer build-up substrate. In general, common build-up substrates can be defined as $n + 2 + n$,

Overall Thickness Options: 260, 360, 560, 610µm

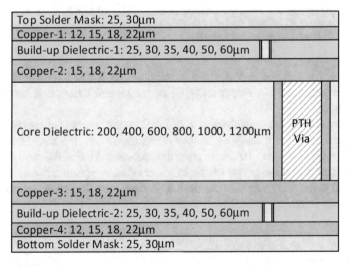

Fig. 4.5 Cross section of a 4-layer laminate substrate. Thickness of each layer is indicated within the layers, depending on the layer thickness selection various total substrate thicknesses such as 260, 360, 560, and 610 µm can be achieved

Fig. 4.6 Cross section of a 1 + 2 + 1 build-up substrate (4-layer) with PTH via through the core and micro-via through the build-up layers, mainly used for high-density flip chip packaging, but can be used in wirebond packaging as well

where n is the number of build-up layer or each side of a 2-layer laminate core. In many designs, no more than 2 layers of laminate core are needed, however, as noted in previous Sect. 4.1, laminate core may consist of number of layers such as 4 or 6 layers. Thus, other variations of build-up substrate with more than 2-laminate layers

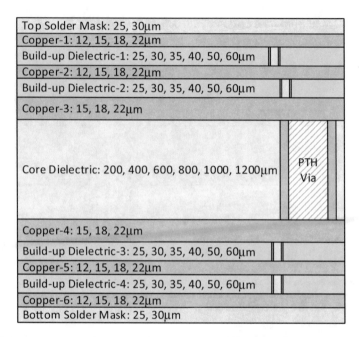

Fig. 4.7 Cross section of a 2 + 2 + 2 build-up substrate (6-layer) with PTH via through the core and micro-via through the build-up layers, mainly used for high-density flip chip packaging, but can be used in wirebond packaging as well

such as 1 + 4 + 1, 3 + 4 + 3, 4 + 6 + 4, are possible. It should be noted that chief routing limitation of flip chip packaging or in general build-up substrate is the laminate core.

As noted earlier, the main drawback of laminate substrates is coarse trace width/ spacing and large via sizes, this limits routing density on the core layer. Design rules for routing on the laminate core are depicted in Fig. 4.8 and Table 4.1. Laminate core thickness is one of the important design parameter that needs to be established at the early stages of the design process, core thickness directly impacts signal/power integrity, reliability, warpage. As shown in Fig. 4.6, core thickness

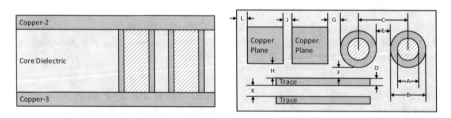

Fig. 4.8 (Left) 2-layer laminate core cross-sectional view. (Right) design parameters for routing and pattern design on laminate core layers, to be used in conjunction with Table 4.1

Table 4.1 Design rule for routing and pattern design on laminate core layers, to be used in conjunction with Fig. 4.8

Core thickness: →		400		800	
		Normal	Advanced	Normal	Advanced
Core process					
A	Minimum PTH drill diameter	300	100	300	150
B	Minimum PTH land diameter	500	200	500	250
C	Minimum PTH pitch	575	250	575	300
D	Minimum trace width	70	50	70	50
Minimum spacing					
E	Via land-to-via land	70	50	70	50
F	Via land-to-trace	70	50	70	50
G	Via land-to-Cu plane	70	50	70	50
H	Trace-to-Cu plane	70	50	70	50
J	Cu plane-to-Cu plane	70	50	70	50
K	Trace-to-trace	70	50	70	50
L	Substrate edge-to-Cu plane spacing	500	500	500	500

can vary from 200 to 1200 μm, other thicknesses are also possible, standard everyday core thickness widely used in the industry is 800 μm, other thicknesses such as 400 μm and less are used to address signal/power integrity requirements. The other criterial factor in selecting a proper core thickness is reliability, especially substrate warpage concern.

The main function of core thickness is to provide mechanical stability and strength to prevent warpage during assembly and package qualification. Depending on the size of substrate a proper core thickness should be selected, obviously for a larger substrate size such as 50 × 50 mm a thicker core is desired. To ensure reliability, a mechanical thermal-stress simulation code can be used to investigate the effect of a selected core thickness on warpage during assembly and temperature cycles. Typically, minimum of 50/50 μm trace width/spacing is standard with 35/45 μm considered more advanced for routing on the core layer, and typical PTH core via land/drill diameter is 250/150 μm with 300 pitch for 800 μm thick core and 240/105 μm with 290 pitch for 400 μm thick core laminate. The fine pitch fan-out and high-density routing capability of build-up substrate are achieved through the build-up layers.

From processing stand point, build-up layers can be patterned with much finer trace width/spacing and much smaller via size compared to laminate substrates, build-up layers are the platform used to fan-out and route high-density interconnects. Build-up substrates are more expensive compared to laminate substrates and are widely used in high performance, high pin count flip chip BGA packages, although it can be used in wirebond packages where certain signal/power integrity performance are required. Common laminate core materials are E705G, 700GR, etc.

Table 4.2 A typical
cross-sectional thicknesses of
a 2 + 2 + 2 build-up substrate

Layers	Thicknesses
Top solder mask	25
Copper-1	15
Build-up dielectric	30
Copper-2	15
Build-up dielectric	30
Copper-3	18
Core dielectric	800
Copper-4	18
Build-up dielectric	30
Copper-5	15
Build-up dielectric	30
Copper-6	15
Bottom solder mask	25

and common build-up materials are GX13, GX92, GY11, GZ41, etc. Typical thicknesses for build-up dielectrics are 25, 30, 35 µm (Figs. 4.6 and 4.7). Vertical routing and transitions among the build-up layers are achieved using micro-via also known as laser-via and build-up-via. As an example, cross section of a 2 + 2 + 2 build-up substrate can be defined as depicted in Table 4.2.

Similar to PTH vias micro-vias also consists of a land pad (capture pad) and a laser drill (Fig. 4.9) but in a much smaller scale compared to PTH vias, typical land/drill dimensions for micro-vias are 100/50, 100/60, 85/60, 80/60, 80/50 µm.

Routing and pattern design on a build-up layer can be performed following the design parameters and specifications illustrated in Fig. 4.10 and Table 4.3.

Various micro-via structures and patterns can be used to escape route within the build-up layers as depicted in Fig. 4.11. Micro-vias from adjacent build-up layers can be stacked on the top of each other to form stacked micro-vias for compact routing, combinations of 2, 3, and 4 micro-vias can be stacked, other options should be investigated from the manufacturer. A non-stacked staggered or simply staggered via pattern can be formed by placing vias from the adjacent layer next to each other so that the via land pads maybe tangent to each other and not overlapped. Combinations of both via patterns can be used to fan-out and maneuver the signals/supplies in the vertical direction.

Fig. 4.9 A build-up via,
consisting of land pad
(capture pad) and laser drill
hole

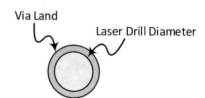

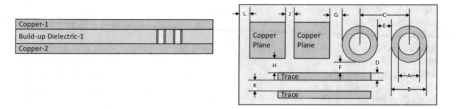

Fig. 4.10 (Left) Cross section of a build-up layer. (Right) design parameters for routing and pattern design on build-up layers, to be used in conjunction with Table 4.3

Table 4.3 Design rule for routing and pattern design on build-up layers, to be used in conjunction with Fig. 4.10

	Build-up process: →		Normal	Advanced
A	Minimum micro-via drill diameter	Non-stacked	50	50
		2 stacked	75	50
		3 stacked	75	75
B	Minimum micro-via land diameter	Non-stacked	100	90
		2 stacked	125	100
		3 stacked	125	125
C	Minimum micro-via pitch	Non-stacked	130	110
		2 stacked	155	120
		3 stacked	155	145
D	Minimum trace width		18 (die area) 25 (global area)	15 (die area) 25 (global area)
Minimum spacing				
E	Via land-to-via land		30	20
F	Via land-to-trace		20	18
G	Via land-to-Cu plane		50	50
H	Trace-to-Cu plane		50	50
J	Cu plane-to-Cu plane		50	50
K	Trace-to-trace		20 (die area) 25 (global area)	15 (die area) 25 (global area)
L	Substrate edge-to-Cu plane		500	500

Advances in substrate processing technologies have made it possible to manu-facture build-up substrates without core, known as core-less substrate, Fig. 4.12. With the absence of core routing limitations, core-less substrate provides much higher routing densities and superior signal/power integrity, however, warpage is a challenge that must be managed.

Fig. 4.11 Various micro-vias structures and patterns can be used to route within the build-up layers. (*A*) 3 micro-vias in staggered configuration, (*B*) another staggered configuration of 3 micro-vias, (*C*) micro-vias from 2 build-up layers are stacked while the third one is staggered, (*D*) all 3 micro-vias are stacked

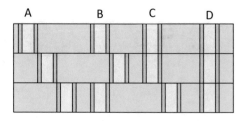

A: Non-Stacked staggered routing
B: Non-Stacked staggered routing
C: 2 Stacked staggered routing
D: 3 Stacked non-staggered routing

Fig. 4.12 A 6-layer core-less substrate with 3 stacked and 2 stacked micro-vias, mainly used for high-density flip chip packaging

Top Solder Mask
Copper-1
Build-up Dielectric-1
Copper-2
Build-up Dielectric-2
Copper-3
Build-up Dielectric-3
Copper-4
Build-up Dielectric-4
Copper-5
Build-up Dielectric-5
Copper-6
Bottom Solder Mask

4.2 Stack-Up and Routing Consideration

Aside from mechanical support and protecting the device, one of the main functions of the substrate is to provide power and signal delivery to the device. Depending on the IP interface, proper transmission line and power delivery structures must be architected to meet the IP interface specifications and requirements. Typical specifications might include certain impedance, delay, cross talk, insertion/reflection loss, or timing budget. Before planning for routing, critical signals must be identified and proper transmission line structures should be implemented.

Given the stack-up nature of the substrate, signal transmission is achieved by means of various transmission line structures, such as micro-strip, embedded micro-strip, and stripline as depicted in Fig. 4.13. Routing on the top or bottom layer of substrate results in micro-strip structure, whereas routing on the internal layers results in stripline and embedded micro-strip structures. In micro-strip structure, the electric fields will propagate through the surrounding materials, including solder mask, air and dielectric material each with a different dielectric constant.

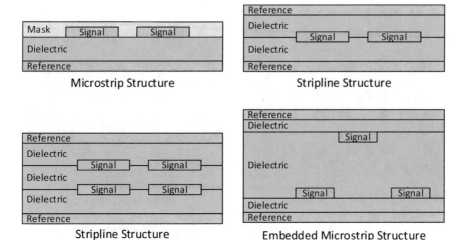

Fig. 4.13 Various transmission line structures

Due to the presence of air in the stack-up, the characteristic impedance of the traces can be higher and with lower frequency dependent losses. On the other hand, in stripline structures entire fields are contained within the dielectric material. A field solver must be used to design a transmission structure for a targeted characteristic impedance.

Routing strategy depends on the device interface specifications and requirements, for example, in routing a DDRx interface, double data rate signals such as DM, DQ, and DQ/DQS should be routed first with strict adherence to budgets (1/4 of a clock period margin for set-up or hold-time relative to strobe), next single data rate signals, differential clock (Address/Command/Control) with looser budget (1/2 of a clock period margin for set-up and hold-time) and, finally, Vref and other support signals such as JTAG.

Many IP interfaces require solid reference planes for good signal/power integrity, cross talk management, and impedance control. As an example, for a memory interface, minimum of 4-layer substrate should be used while 6 or more layers are preferred, moreover, in order to manage the cross talk and improve signal/power integrity a build-up substrate should be used. In a 4-layer stack-up implementation as depicted in Fig. 4.14, signals are routed as micro-strip while the internal planes provide the solid reference planes for power and ground. In a 6-layer stack-up implementation as illustrated in Fig. 4.15, signals are routed as micro-strip on layers 1 and 6, with additional signals routed as stripline on layers 3 and 4 in order to ease the routing congestion (cross talk control) between the DDRx controller–PHY interface and memory device. Internal planes 2 and 5 are allocated as ground to provide solid clean references for signals on layers 1 and 3 as well as 4 and 6. Power domains are allocated on layers 3 and 4 to power, this creates tight coupling between the power and ground planes which results in lower power inductance.

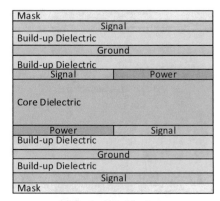

Fig. 4.14 Typical 4 and 6-layers build-up substrate stack-up for DDRx interface routing

It should be noted that different DDRx signal groups have different loading configurations, thus requiring a different routing strategy. In order to minimized the routing layers and lower the substrate cost, controller/PHY bump pattern must be optimized with respect to the memory device. Since the memory device ball pattern is fixed by the memory supplier, an optimal controller/PHY bump pattern can minimize the trace crossover; this not only minimizes the number of layers and lowers the cost but also improves the performance. This brings us to the concept of pathfinding and co-design that will be discussed in more detail in the following chapters.

4.3   Silicon Substrate (Interposer)

Limitations in organic substrate manufacturing processes have prompted the industry to use silicon substrate also known as silicon interposer to fan-out very fine pitch devices. For devices with finer bump pitch, typically less than 80 μm a silicon interposer can be used to fan-out the fine bump pitch to a coarser pitch. Silicon interposer has been effectively used in the industry to fan-out 40 μm bump pitch on the front side to 150 μm bump pitch on the backside. Subsequently, an organic build-up substrate is used to fan-out 150 μm to package BGA ball pitch such as 400, 600, 800 μm, and 1 mm. We will first study the silicon interposer manufacturing process.

Processing steps used to manufacture silicon interposer are depicted in Fig. 4.15. Silicon interposer manufacturing process involves six distinct steps as illustrated:

1. Via Formation: To form vias, a full thickness blank silicon wafer is patterned and etched to create via holes of certain diameter and depth, followed by depositing barrier layer (oxide layer) and copper fill process. A chemical

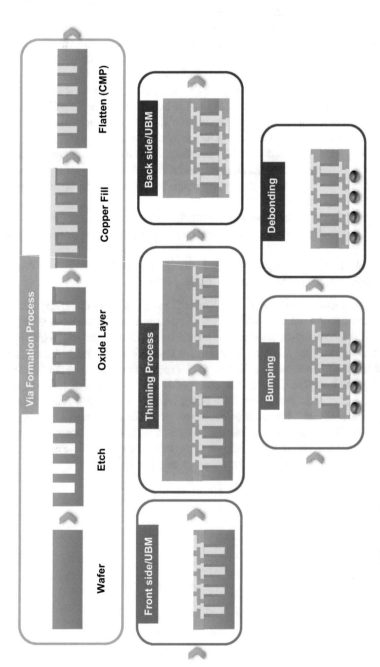

Fig. 4.15   Silicon interposer fabrication process

mechanical polishing/planarization (CMP) process is used to remove the excess copper to create isolated blind vias.

2. Front side redistribution layer (RDL) and under bump metallization (UBM) formation: Once the isolated vias are formed, front side RDL (wiring) and under bump metallization (UBM) are formed. The UBM corresponds to device fine bump pitch and pattern, i.e., 40 μm pitch.

3. Thinning process: In order to expose the blind vias, wafer must be thinned. To thin the wafer a carrier (silicon or glass) is bonded to the wafer, this process is known as bonding process. Wafer is then thinned to expose the via. At this stage, the exposed via is known as through silicon via (TSV).

4. Backside RDL and under bump metallization (UBM) formation: While the carrier is still attached to the thinned wafer, backside RDL (wiring), and UBM are formed. The UBM typically corresponds to flip chip bump pitch and pattern, i.e., 150 μm pitch.

5. Bumping: UBM formed in step 4 provides a landing pad to bump the wafer. Depending on the pitch required either cupper pillar or solder bumps are used to bump the wafer.

6. Debonding: The last stage involves removing the carrier wafer from the structure, this process is known as debonding process.

Silicon manufacturing processes have made it possible to produce very fine trace width/spacing in the order of sub-micron. With copper damascenes processes, RDL of 0.1 μm line/spacing or smaller can be achieved. One of the key parameters in design and fabrication of silicon interposer is the TSV aspect ratio. Aspect ratio is referred to as via diameter with respect to interposer thickness (via depth). Typical industry grade interposer has TSV diameter of 10 μm and depth of 100 μm, known as 10:100 aspect ratio interposer. Depending on the FAB process, capabilities and expertize other aspect ratios such as 20:200 are possible. From signal integrity point of view, it should be noted that due to lossy nature of silicon, transmitting high-speed signals through the silicon interposer TSV is a bit challenging and requires other solutions beyond the scope of this book.

Chapter 5
Conventional Design Flow

Abstract In this chapter, we will introduce the wirebond and flip chip package design process and methodology. First, we will review the conventional design flow followed by detailed walk through examples to demonstrate the design process. Design and methodology presented in this chapter builds the foundation of modular integration.

5.1 Traditional Package Design Flow

Package design flow has evolved considerably over the years primarily due to emerging markets and applications. Traditionally, a device is designed first followed by substrate design and assembly process. Such flow as depicted in Fig. 5.1 is known in the industry as "Through-Over-the-Wall" flow. As evident, this flow is similar to a one-way street, designed devices are handed over to the packaging team for packaging and assembly, and there are minimal or no interaction between the die design team and package design team. Such flow results in over designed package and more expensive product as a consequence.

A more descriptive version of Through-Over-the-Wall flow is illustrated in Fig. 5.2. As can be seen, the entire flow is one directional. Final device specification is handed over to the packaging team, and substrate design team will then use the substrate design rules presented in Chap. 4 to define the substrate cross section followed by device placement, wirebonding or bump fan-out, routing, verification, documentation, and finally generating manufacturing data such as GERBER, DXF, GDSII. At the request of some customers, thermal simulation data, RLC parasitic, s-parameters data may be generated also known as performance data.

© Springer International Publishing AG 2018   53
F. Yazdani, *Foundations of Heterogeneous Integration: An Industry-Based,
2.5D/3D Pathfinding and Co-Design Approach*,
https://doi.org/10.1007/978-3-319-75769-8_5

Fig. 5.1 Traditional one-directional "Through-Over-the-Wall" design flow

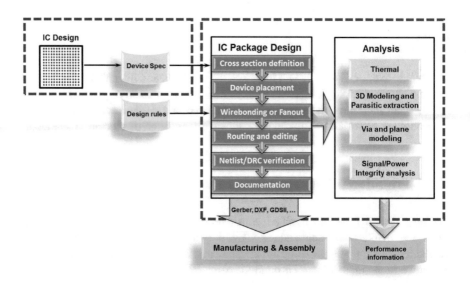

Fig. 5.2 A detailed version of Through-Over-the-Wall package design flow

5.2   Substrate Design

From package design perspective, our main focus is on substrate design. As noted in previous chapters, substrate design has direct impact on overall product cost and performance; thus, substrate design is the most critical step in any product development. It is assumed that the reader has thoroughly mastered the substrate technologies and design rules presented in Chap. 4. In this section, we will focus on substrate design flow and methodology.

5.2.1   Library Development Process

Every design requires a set of building blocks (elements) stored in a folder called library. An essential part of the substrate design process is library development for all the elements to be used in the design. The very first step in the design process is to define a library of elements, and the library can be modified for reuse in subsequent designs as well. Normally, the library consists of padstacks for various via sizes and configurations (blind, buried, through hole), various pad sizes based on device

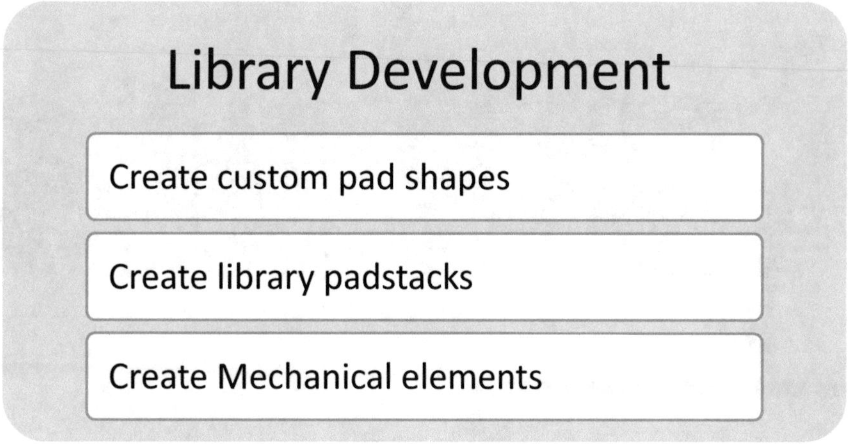

Fig. 5.3 Typical library development elements

configuration and substrate BGA requirements, typical library development element is illustrated in Fig. 5.3. Other elements in the library are device file containing logic and a constraint file containing specific design rules (physical, electrical) and cross-sectional information that can be modified and reused in another design.

5.2.2 Define Substrate

A substrate is defined by its cross section. Cross-sectional information includes order of layer stack-up, dielectric materials and thicknesses, metal materials and thicknesses, electrical behavior, and shielding (Fig. 5.4).

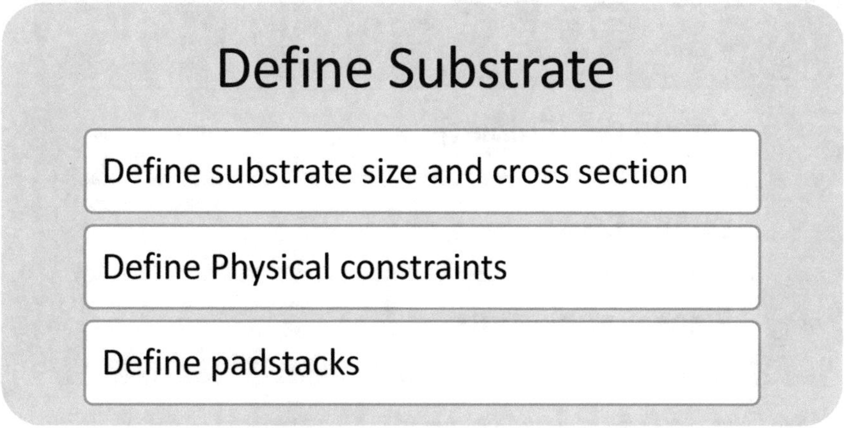

Fig. 5.4 Substrate definition information

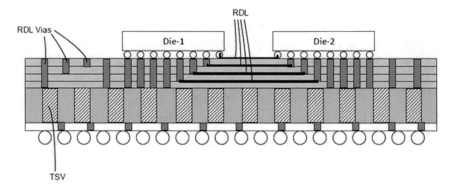

Fig. 5.5 A typical interposer cross section

As an example, a silicon interposer substrate cross section is illustrated in Fig. 5.5. Cross-sectional information and libraries should include bulk silicon thickness, RDL metal thicknesses, dielectric materials (i.e., silicon oxide, poly-imide) and thickness, RDL via size and type (blind, buried, etc.), TSV size, chiplet landing pads, and C4 pad sizes.

5.2.3 Define Components

Figure 5.6 illustrates the elements of a package that must be defined. A device can be defined in various formats such as Lef/Def, and this data format provides more detailed information about the die and its underlying elements such as I/O cells, macro-size, and location. Exchanging this type of data is more confined within an

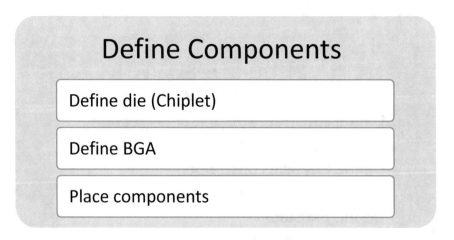

Fig. 5.6 Components such as die and substrate BGA ball pattern must be defined

Table 5.1 Device
information in coordinate
format

Pin number	X coord	Y coord	Net name
1	−4075.2	0	VSS
2	−4255.2	0	DQ1A
3	−4435.2	0	VSS
4	−4615.2	0	VDD

organization that is performing the die design and is more difficult and restricted to be shared with outside vendors for many legal reasons.

- Design Exchange Format (DEF): Provides physical aspects of the device, which include netlist and cell location.
- Library Exchange Format (LEF): Provides cell library, which includes cell location, macro-/block location, size, and connectivity.

The most common format to exchange device information whether with internal teams or external organizations is the text coordinate format. IC design team normally provides the device information in the form of a text file or a spreadsheet. Table 5.1 illustrates the format used to communicate device information. The text file contains pin number (bump/pad number), pin (bump/pad) location (x, y coordinate), and netlist (logic) associated with each pin (bump/pad). Likewise, the substrate BGA balls or interposer C4 bump location is communicated similar to the device format. We will take a closer look at this in the next chapter.

5.2.4   Define Connectivity

Figure 5.7 illustrates the connectivity elements that need to be defined. The main objective to design an optimal substrate is to define an optimal connectivity.

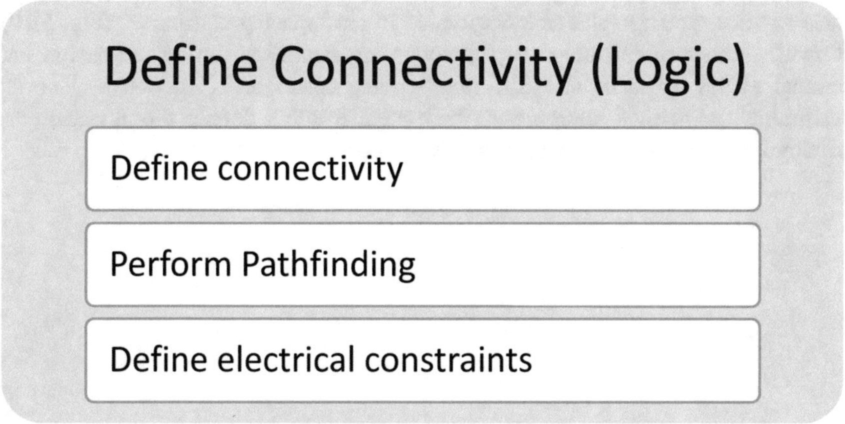

Define Connectivity (Logic)

Define connectivity

Perform Pathfinding

Define electrical constraints

Fig. 5.7 Typical steps to establish optimal connectivity

A connectivity netlist that defines the connection between the die and package BGA ball must be established. This connectivity netlist must be optimized to provide optimal routing between the device and substrate BGA balls; in addition, it must provide optimal signal/power integrity performance. We define the process of optimizing the connectivity netlist as "Pathfinding." Connectivity and pathfinding among the devices and BGA or interposer C4 bump play a critical role in system performance, routing, and overall costs.

It is important to realize that the package substrate design is primarily governed by the device bump/pad pattern and pitch and in many cases governed by the substrate BGA ball pattern and pitch also known as package ball-out. We will study pathfinding in more detail in the next chapter.

5.2.5   Define Physical Implementation

Elements and steps to define physical implementation are illustrated in Fig. 5.8. Constraint is a user-defined rule that drives connectivity and routing. In order to perform physical implementation of the substrate, there are various types of constraint/rule that must be defined such as spacing, physical, and electrical rules.

Spacing constraint: Defines the spacing between different elements in the design. Figure 5.9 illustrates spacing constraints applied to vias and traces during escape routing from a flip chip die. Some of the most important and commonly used spacing constraints for single-ended as well diff pairs are illustrated in Fig. 5.10.

Physical constraint: Defines the constraints that govern individual nets and includes trace width on each layer and includes differential pairs trace width/spacing.

Electrical constraint: Defines the rules that govern the electrical behavior of individual net, differential pairs, or a bus. Timing delay constraint, for example, is normally specified for nets or group of nets (bus) to a specific allowable delay in a driver/receiver path. Maximum cross talk constraint specifies the maximum allowable cross talk on a victim trace from all aggressor nets. Normally, cross talk is simulated during per-design analysis using an EM tool and the optimal trace-to-trace spacing value is incorporated in the "spacing constraint" (Fig. 5.10). Similarly, other signal integrity parameters are simulated during pre-design and optimal values are incorporated in the "spacing constraint" (Fig. 5.10). Once the constraints are defined, design must be routed. Routing defines the physical connection between the die and package BGA balls.

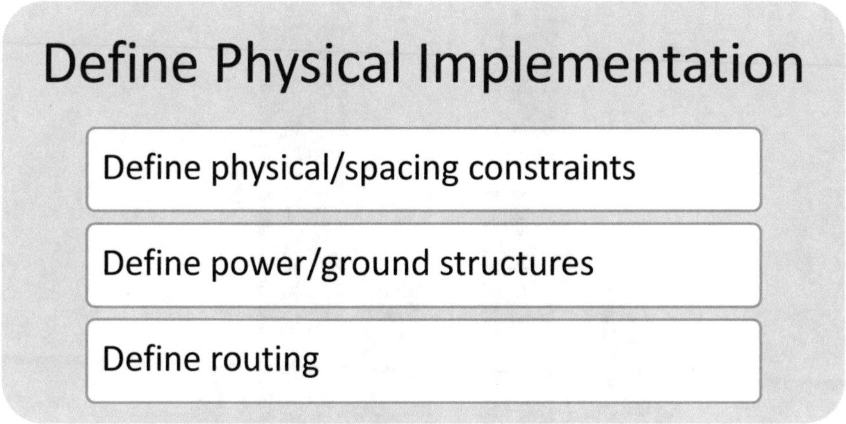

Define Physical Implementation

Define physical/spacing constraints

Define power/ground structures

Define routing

Fig. 5.8 Elements and steps to perform physical implementation

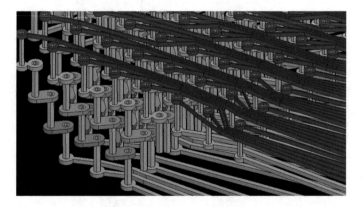

Fig. 5.9 Example of under the die escape routing based on spacing constraint, via-to-via, trace-to-trace, and trace-to-via

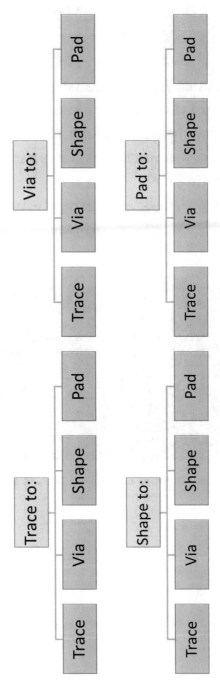

Fig. 5.10 Example of required spacing constraint to perform physical implementation

5.2.6 Verification

All designs must be verified to ensure adherence to manufacturing specifications as well as performance criteria. Figure 5.11 illustrates common verification steps. Before a design can be released to manufacturing a series of verification steps must be performed. Connectivity must be verified to make sure proper connection between the die pins and package balls, this includes all the signals and supply nets. A comprehensive design rule check (DRC) must be performed to resolve any violation either physical, electrical, mechanical, etc. Finally, to make sure that the design meets the performance criteria a comprehensive signal/power integrity analysis as well as thermal and stress analysis must be performed.

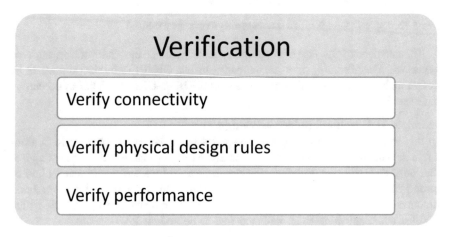

Fig. 5.11 Common verification steps to ensure adherence to manufacturing rules as well as performance target

5.2.7 Reports and Manufacturing Files

Upon completion of the verification process, a complete set of reports must be generated. This typically includes important aspect of the design and specifications, such as conductor length report, time delay report, bill of material report, netlist report, cross-sectional report (Fig. 5.12).

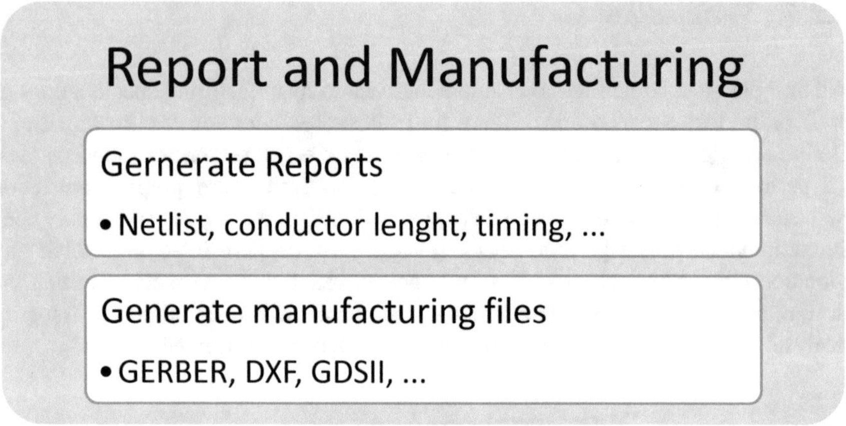

Report and Manufacturing

Gernerate Reports

• Netlist, conductor lenght, timing, …

Generate manufacturing files

• GERBER, DXF, GDSII, …

Fig. 5.12 Typical report files and manufacturing files to be generated

To manufacture the designed substrate, manufacturing file and drawing must be generated. Manufacturing files are primarily in the GERBER format, and all EDA design tools provide the option of generating GERBER, DXF, or GDSII as needed by the manufacturer.

Example 1: Wirebond Substrate Design

In this example, we will design a low-cost 4-layer wirebond substrate to demonstrate the substrate design process. Typically, a low-cost laminate substrate requires that only PTH via type to be used in the design, and we will avoid using blind or buried via type in this example. We will place and route a 2 mm × 2 mm wirebond die on a 10 mm × 10 mm BGA substrate with 0.8-mm BGA ball pitch. Die netlist is fixed while substrate BGA balls are free to be assigned

Step-1: Set up substrate cross section

Set up substrate cross section with 4 layers. Typically, number of layers are determined per routing density and signal/power integrity requirement. In this example, we will use 4 layers to route all the signals and supply nets. Layer stack-up is shown in Table 5.2, and signals can be routed in micro-strip format on layer-1 with reference to ground plane underneath.

Dielectric and copper thicknesses can be varied to meet the signal impedance requirement. Consider 50-Ω impedance requirement for all signals, and use an impedance calculator to determine the trace width necessary to achieve 50-Ω impedance per selected dielectric and copper thicknesses.

Table 5.2 Substrate layer stack-up arrangement

Layer-1	Signal
Layer-2	Ground (vss)
Layer-3	Power (vdd)
Layer-4	Signal

Table 5.3 Padstacks definition

Padstacks	Specification (unit: μm)
Bondfinger	300 × 90 (oblong shape)
Die pad opening	50 × 50 (square shape)
BGA landing pad	450 (circle)
BGA pad opening	350 (circle)
PTH via	300 pad, 150 drill (circle)

Step-2: Set up design rules

Design rules/constraints must be step up for each substrate layer before proceeding with the design. As noted earlier, these rules include manufacturing, electrical, mechanical. Substrate manufacturing design rules were discussed in Chap. 4. For this example, critical spacing constraints to be set up are trace-to-trace, trace-to-via, trace-to-shape, trace-to-bondfinger, via-to-via, via-to-shape, shape-to-shape, via-to-bondfinger, and bondfinger-to-bondfinger.

Step-3: Set up padstacks

Padstacks are the building blocks of the substrate, and we will be using padstacks to define, route, and design the substrate. Padstacks should be created for appropriate substrate layer; for example, BGA landing padstack is created on layer-4, whereas bondfinger padstack is created on layer-1 and PTH via. For this example, create the padstacks according to Table 5.3.

Step-4: Generate BGA Grid

Generate a 10 mm × 10 mm BGA grid with 0.8-mm ball pitch using the BGA padstack created in step 1 (Fig. 5.13). Verify that BGA pad is placed on layer-4. A JEDEC standard 10 mm × 10 mm body size with 0.8-mm ball pitch should result in 144 balls. Note the position A1 ball corner and rows/column numbering. Alternatively, BGA ball coordinates provided in Table 5.4 can be used to generate the substrate footprint.

Step-5: Generate die

Use the die pad opening coordinates provide in Table 5.5 to generate a 2 mm × 2 mm wirebond die. Die pad opening padstack created in step-3 can be used to generate the die (Fig. 5.14). Note that each die pin number is associated with a x, y coordinate and a net name. Normally, power pins are designated using red color, while green color is used to designate ground pins. To avoid bonding out a specific die pad, net name associate with a die pin may be removed. Place the die at the

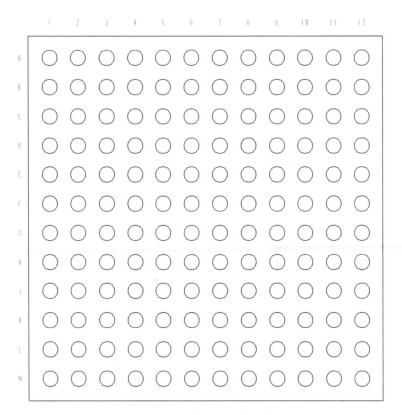

Fig. 5.13 A 10 mm × 10 mm BGA grid with 0.8-mm ball pitch

center of substrate, and note that die is mounted on substrate layer-1. Figure 5.15
shows the die mounted on the substrate.

Step-6: Perform connectivity/routing

In this step, we will generate a set of ground and power rings around the die on
substrate layer-1. Device power and ground pins will be wirebonded to these rings.
Power and ground rings will then be connected to power and ground planes using
PTH vias. Use the guideline and design rules provided in Chap. 4 to generate a set
of ground/power rings with width of 100 μm and spacing of 100 μm in between
rings (Fig. 5.16). Depending on the power integrity requirements, designer will
decide the order of supply rings; typically ground ring is the closest to the device
followed by power rings such as vdd-I/O and vdd-core. Note that rings can be split
into multiple segments to accommodate various supply nets. Note that die supply
pins may be bonded to individual bondfingers and routed individually.

Table 5.4 10 mm × 10 mm
substrate BGA ball coordinate

Pin number	X coord	Y coord
A1	−4400	4400
A2	−3600	4400
A3	−2800	4400
A4	−2000	4400
A5	−1200	4400
A6	−400	4400
A7	400	4400
A8	1200	4400
A9	2000	4400
A10	2800	4400
A11	3600	4400
A12	4400	4400
B1	−4400	3600
B2	−3600	3600
B3	−2800	3600
B4	−2000	3600
B5	−1200	3600
B6	−400	3600
B7	400	3600
B8	1200	3600
B9	2000	3600
B10	2800	3600
B11	3600	3600
B12	4400	3600
C1	−4400	2800
C2	−3600	2800
C3	−2800	2800
C4	−2000	2800
C5	−1200	2800
C6	−400	2800
C7	400	2800
C8	1200	2800
C9	2000	2800
C10	2800	2800
C11	3600	2800
C12	4400	2800
D1	−4400	2000
D2	−3600	2000

(continued)

Pin number	X coord	Y coord
D3	−2800	2000
D4	−2000	2000
D5	−1200	2000
D6	−400	2000
D7	400	2000
D8	1200	2000
D9	2000	2000
D10	2800	2000
D11	3600	2000
D12	4400	2000
E1	−4400	1200
E2	−3600	1200
E3	−2800	1200
E4	−2000	1200
E5	−1200	1200
E6	−400	1200
E7	400	1200
E8	1200	1200
E9	2000	1200
E10	2800	1200
E11	3600	1200
E12	4400	1200
F1	−4400	400
F2	−3600	400
F3	−2800	400
F4	−2000	400
F5	−1200	400
F6	−400	400
F7	400	400
F8	1200	400
F9	2000	400
F10	2800	400
F11	3600	400
F12	4400	400
G1	−4400	−400
G2	−3600	−400
G3	−2800	−400
G4	−2000	−400

(continued)

Table 5.4 (continued)

Pin number	X coord	Y coord
G5	−1200	−400
G6	−400	−400
G7	400	−400
G8	1200	−400
G9	2000	−400
G10	2800	−400
G11	3600	−400
G12	4400	−400
H1	−4400	−1200
H2	−3600	−1200
H3	−2800	−1200
H4	−2000	−1200
H5	−1200	−1200
H6	−400	−1200
H7	400	−1200
H8	1200	−1200
H9	2000	−1200
H10	2800	−1200
H11	3600	−1200
H12	4400	−1200
J1	−4400	−2000
J2	−3600	−2000
J3	−2800	−2000
J4	−2000	−2000
J5	−1200	−2000
J6	−400	−2000
J7	400	−2000
J8	1200	−2000
J9	2000	−2000
J10	2800	−2000
J11	3600	−2000
J12	4400	−2000
K1	−4400	−2800
K2	−3600	−2800
K3	−2800	−2800
K4	−2000	−2800
K5	−1200	−2800
K6	−400	−2800

(continued)

Table 5.4 (continued)

Pin number	X coord	Y coord
K7	400	−2800
K8	1200	−2800
K9	2000	−2800
K10	2800	−2800
K11	3600	−2800
K12	4400	−2800
L1	−4400	−3600
L2	−3600	−3600
L3	−2800	−3600
L4	−2000	−3600
L5	−1200	−3600
L6	−400	−3600
L7	400	−3600
L8	1200	−3600
L9	2000	−3600
L10	2800	−3600
L11	3600	−3600
L12	4400	−3600
M1	−4400	−4400
M2	−3600	−4400
M3	−2800	−4400
M4	−2000	−4400
M5	−1200	−4400
M6	−400	−4400
M7	400	−4400
M8	1200	−4400
M9	2000	−4400
M10	2800	−4400
M11	3600	−4400
M12	4400	−4400

Next wirebond guard rings are created on the periphery of the rings as illustrated in Fig. 5.17. Guard rings are used to position the bondfingers around the outer periphery of the rings. Guard rings can be of any shape and length and can be placed at any distance from the rings. However, to minimize the wire length, guard rings should be placed at minimal distance from the rings without violating any design rule, and normally number of iterations are involved to minimize the wire lengths. As a general rule, signals are bonded to bondfingers and supply nets are bonded to rings.

Table 5.5 2 mm × 2 mm wirebond die pad opening coordinate

Pin number	X coord	Y coord	Net name
1	−850	950	SIG_1
2	−750	950	VDD
3	−650	950	SIG_3
4	−550	950	SIG_4
5	−450	950	VSS
6	−350	950	SIG_6
7	−250	950	SIG_7
8	−150	950	VDD
9	−50	950	SIG_9
10	50	950	SIG_10
11	150	950	VSS
12	250	950	SIG_12
13	350	950	SIG_13
14	450	950	VDD
15	550	950	SIG_15
16	650	950	SIG_16
17	750	950	VSS
18	850	950	SIG_18
19	950	850	VDD
20	950	750	SIG_20
21	950	650	SIG_21
22	950	550	VSS
23	950	450	SIG_23
24	950	350	SIG_24
25	950	250	VDD
26	950	150	SIG_26
27	950	50	SIG_27
28	950	−50	VSS
29	950	−150	SIG_29
30	950	−250	SIG_30
31	950	−350	VDD
32	950	−450	SIG_32
33	950	−550	SIG_33
34	950	−650	VSS
35	950	−750	SIG_35
36	950	−850	SIG_36
37	850	−950	SIG_37
38	750	−950	SIG_38
39	650	−950	VSS

(continued)

Table 5.5 (continued)

Pin number	X coord	Y coord	Net name
40	550	−950	SIG_40
41	450	−950	SIG_41
42	350	−950	VDD
43	250	−950	SIG_43
44	150	−950	SIG_44
45	50	−950	VSS
46	−50	−950	SIG_46
47	−150	−950	SIG_47
48	−250	−950	VDD
49	−350	−950	SIG_49
50	−450	−950	SIG_50
51	−550	−950	VSS
52	−650	−950	SIG_52
53	−750	−950	SIG_53
54	−850	−950	VDD
55	−950	−850	SIG_55
56	−950	−750	VSS
57	−950	−650	SIG_57
58	−950	−550	SIG_58
59	−950	−450	VDD
60	−950	−350	SIG_60
61	−950	−250	SIG_61
62	−950	−150	VSS
63	−950	−50	SIG_63
64	−950	50	SIG_64
65	−950	150	VDD
66	−950	250	SIG_66
67	−950	350	SIG_67
68	−950	450	VSS
69	−950	550	SIG_69
70	−950	650	SIG_70
71	−950	750	VDD
72	−950	850	SIG_72

Fig. 5.14 A 2 mm × 2 mm wirebond die generated from netlist in Table 5.5

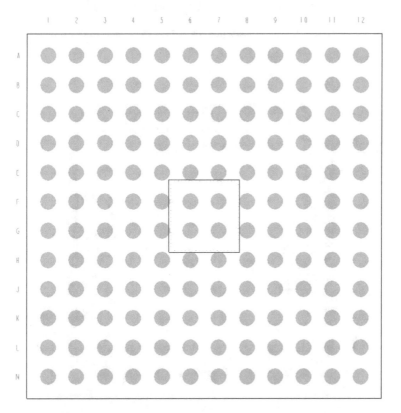

Fig. 5.15 Die mounted on layer-1 and positioned at the center of the substrate

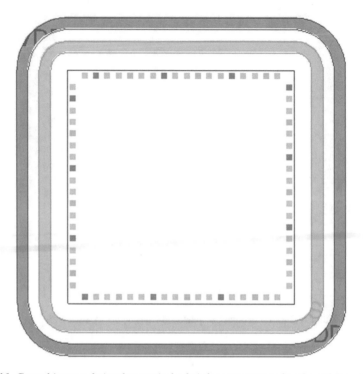

Fig. 5.16 Ground (green color) and power (red color) rings are generated on the periphery of the die

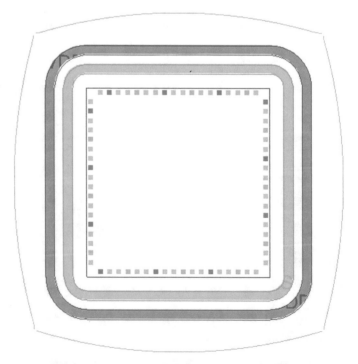

Fig. 5.17 Wirebond guard rings are placed on the outer periphery of the ring

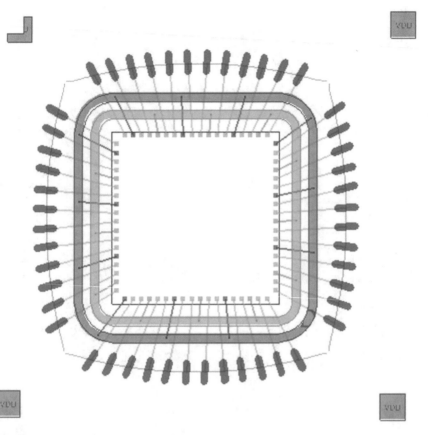

Fig. 5.18 Device signal pins are wirebonded to bondfingers while power/ground pins are bonded to the rings

Position the bondfingers on the guard rings as illustrated in Fig. 5.18. Device signal pins are wirebonded to bondfingers, while power/ground pins are bonded to the rings. At this point, fiducial markers can be placed on four corners of the die/ substrate; a unique fiducial shape is used to identify the package A1 ball; it is very important to be able to identify the package A1 ball corner; in this example, an inverted L shape is used to identify the A1 ball corner as illustrated in Fig. 5.18.

Upon completion of the wirebonding process, connectivity between the die pins and substrate BGA balls must be established. Since the die netlist is fixed, we cannot swap or change any pin on the die; however, BGA balls are free to be assigned; in this flow, die is deriving the connectivity. We will start this process by first assigning power and ground nets to the BGA balls, depending on the signal/

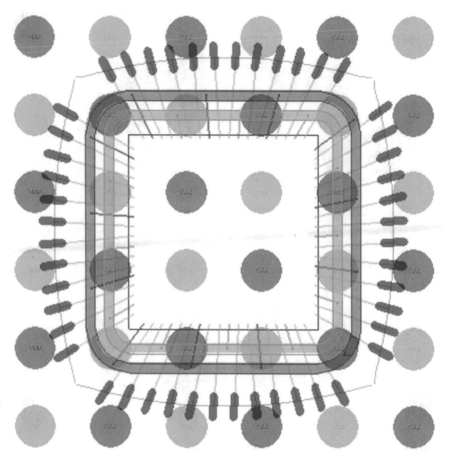

Fig. 5.19 Power and ground BGA assignment for balls right under the die

power integrity requirements it is good to establish a supply-to-signal ratio such as 1-to-3, implying 1 supply ball (power or ground net) is needed for every 3 signal balls.

Start by assigning power and ground nets to BGA balls right under the die in a checkerboard format as illustrated in Fig. 5.19. Signals are normally assigned to the outer rows of substrate; this would facilitate signal escape routing on the PCB.

Similarly, assign power and ground nets in checkerboard format to BGA balls located on 4 corners of the substrate; each group consists of 9 balls as illustrated in Fig. 5.20.

Fig. 5.20 Assign power and ground nets in checkerboard format to BGA balls located on 4 corners of the substrate

There are 4 quadrants, each with 18 BGA balls unassigned. Given there are 12 bondfinger signal nets on each side of the die, proceed by assigning power and ground nets to 6 balls on north side of the substrate followed by assigning signals to rest of the remaining balls as illustrated in Fig. 5.21. From routing perspective, to make sure we do not encounter crossover issues during routing make sure ratnets do not crossover. Ratnets are used to identify assigned pins and to assess routability. Repeat the same ball assignment pattern for other quadrants, as illustrated in Fig. 5.22. Table 5.6 provides the BGA ball coordinates with net names fully assigned.

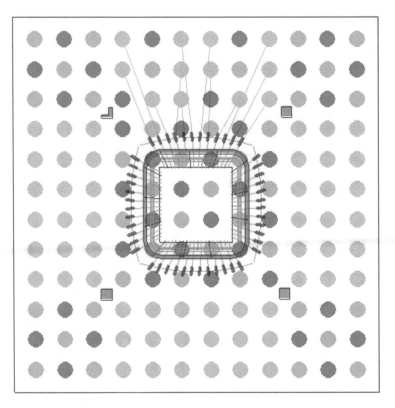

Fig. 5.21 Power/ground and signal are assigned to BGA balls on north side of the substrate with no ratnets crossover

We are now in a position to route the design and, there are number of approaches and strategies to route the design; however, we will start by dropping a PTH via for each BGA ball followed by dropping vias for power and ground rings as illustrated in Fig. 5.23.

Remove the PTH via associated with A1 ball corner, and connect the A1 ball to nearby ball or via as illustrated in Fig. 5.24. Place a right triangle fiducial in the corner to signify the package A1 ball corner, and note that this fiducial should be placed on layer-4 (Fig. 5.24).

Next we will add the power and ground planes to the design. Note that there should be a clearance between the planes and substrate edge, normally in the range of 300–500 μm. Add a ground plane to layer-2, and at this point, all PTH ground vias should be shorted to ground plane, while power and signal vias should be

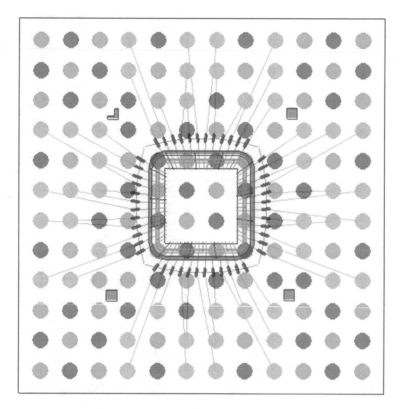

Fig. 5.22 Completed package BGA ball assignment with no ratnets crossover

voided (Fig. 5.25). Similarly, add a power plane to layer-3, and all PTH power vias should be shorted to power plane, while ground and signal vias should be voided (Fig. 5.26).

Manually or automatically route the design by connecting bondfingers to PTH vias on layer-1. Your design should be fully routed at this point, and substrate top view (layer-1) illustrated in Figs. 5.27 and 5.28 shows the final design with all layers visible.

Step: 7: Verification and manufacturing

We are now at the most critical stage of the design process, and the entire design must be verified for any design rule violation and any signal/power integrity violation. Run a comprehensive design rule check on the design, if any DRC violation is observed, identify the cause of the violation and try to resolve it. If the DRC

Table 5.6 10 mm × 10 mm
substrate BGA ball
coordinates with net names
assigned

Pin number	X coord	Y coord	Net name
A1	−4400	4400	VSS
A2	−3600	4400	VDD
A3	−2800	4400	VSS
A4	−2000	4400	SIG_3
A5	−1200	4400	VDD
A6	−400	4400	SIG_9
A7	400	4400	SIG_10
A8	1200	4400	VDD
A9	2000	4400	SIG_15
A10	2800	4400	VSS
A11	3600	4400	VDD
A12	4400	4400	VSS
B1	−4400	3600	VDD
B2	−3600	3600	VSS
B3	−2800	3600	VDD
B4	−2000	3600	SIG_1
B5	−1200	3600	VSS
B6	−400	3600	SIG_6
B7	400	3600	SIG_12
B8	1200	3600	VSS
B9	2000	3600	SIG_16
B10	2800	3600	VDD
B11	3600	3600	VSS
B12	4400	3600	VDD
C1	−4400	2800	VSS
C2	−3600	2800	VDD
C3	−2800	2800	VSS
C4	−2000	2800	VDD
C5	−1200	2800	SIG_4
C6	−400	2800	SIG_7
C7	400	2800	VDD
C8	1200	2800	SIG_13
C9	2000	2800	SIG_18
C10	2800	2800	VSS
C11	3600	2800	VDD
C12	4400	2800	VSS
D1	−4400	2000	SIG_67
D2	−3600	2000	SIG_70

(continued)

Table 5.6 (continued)

Pin number	X coord	Y coord	Net name
D3	−2800	2000	SIG_72
D4	−2000	2000	VDD
D5	−1200	2000	VSS
D6	−400	2000	VDD
D7	400	2000	VSS
D8	1200	2000	VDD
D9	2000	2000	VSS
D10	2800	2000	SIG_20
D11	3600	2000	SIG_21
D12	4400	2000	SIG_24
E1	−4400	1200	VDD
E2	−3600	1200	VSS
E3	−2800	1200	SIG_69
E4	−2000	1200	VSS
E5	−1200	1200	VDD
E6	−400	1200	VSS
E7	400	1200	VDD
E8	1200	1200	VSS
E9	2000	1200	VDD
E10	2800	1200	SIG_23
E11	3600	1200	VSS
E12	4400	1200	VDD
F1	−4400	400	SIG_63
F2	−3600	400	SIG_64
F3	−2800	400	SIG_66
F4	−2000	400	VDD
F5	−1200	400	VSS
F6	−400	400	VDD
F7	400	400	VSS
F8	1200	400	VDD
F9	2000	400	VSS
F10	2800	400	VDD
F11	3600	400	SIG_26
F12	4400	400	SIG_27
G1	−4400	−400	SIG_60
G2	−3600	−400	SIG_61
G3	−2800	−400	VDD
G4	−2000	−400	VSS

(continued)

Table 5.6 (continued)

Pin number	X coord	Y coord	Net name
G5	−1200	−400	VDD
G6	−400	−400	VSS
G7	400	−400	VDD
G8	1200	−400	VSS
G9	2000	−400	VDD
G10	2800	−400	SIG_30
G11	3600	−400	SIG_32
G12	4400	−400	SIG_29
H1	−4400	−1200	VDD
H2	−3600	−1200	VSS
H3	−2800	−1200	SIG_58
H4	−2000	−1200	VDD
H5	−1200	−1200	VSS
H6	−400	−1200	VDD
H7	400	−1200	VSS
H8	1200	−1200	VDD
H9	2000	−1200	VSS
H10	2800	−1200	SIG_33
H11	3600	−1200	VSS
H12	4400	−1200	VDD
J1	−4400	−2000	SIG_57
J2	−3600	−2000	SIG_55
J3	−2800	−2000	VSS
J4	−2000	−2000	VSS
J5	−1200	−2000	VDD
J6	−400	−2000	VSS
J7	400	−2000	VDD
J8	1200	−2000	VSS
J9	2000	−2000	VDD
J10	2800	−2000	VDD
J11	3600	−2000	SIG_36
J12	4400	−2000	SIG_35
K1	−4400	−2800	VSS
K2	−3600	−2800	VDD
K3	−2800	−2800	VSS
K4	−2000	−2800	VDD
K5	−1200	−2800	SIG_50
K6	−400	−2800	VDD

(continued)

Table 5.6 (continued)

Pin number	X coord	Y coord	Net name
K7	400	−2800	SIG_43
K8	1200	−2800	SIG_40
K9	2000	−2800	SIG_37
K10	2800	−2800	VSS
K11	3600	−2800	VDD
K12	4400	−2800	VSS
L1	−4400	−3600	VDD
L2	−3600	−3600	VSS
L3	−2800	−3600	VDD
L4	−2000	−3600	SIG_53
L5	−1200	−3600	VSS
L6	−400	−3600	SIG_47
L7	400	−3600	SIG_44
L8	1200	−3600	VSS
L9	2000	−3600	SIG_38
L10	2800	−3600	VDD
L11	3600	−3600	VSS
L12	4400	−3600	VDD
M1	−4400	−4400	VSS
M2	−3600	−4400	VDD
M3	−2800	−4400	VSS
M4	−2000	−4400	SIG_52
M5	−1200	−4400	VDD
M6	−400	−4400	SIG_49
M7	400	−4400	SIG_46
M8	1200	−4400	VDD
M9	2000	−4400	SIG_41
M10	2800	−4400	VSS
M11	3600	−4400	VDD
M12	4400	−4400	VSS

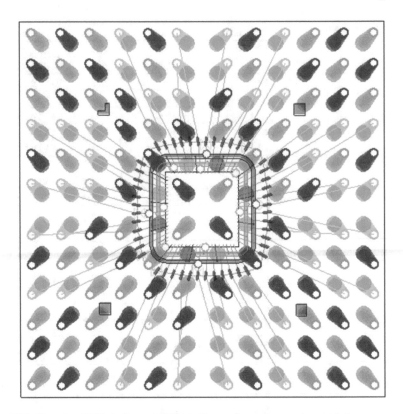

Fig. 5.23 Dropping PTH via for each BGA ball as well as power and ground rings

violation is due to spacing constraint, try to slide the elements to resolve the violation. If, for example, the violation is due to timing error, vary the conductor length to resolve the timing issue. Once the design is free of any DRC violation, generate a set of report on important criteria such as conductor length, timing, and die-to-package netlist report; these reports are mandatory and required for sign-off and release the design to manufacturing. All tools provide capabilities to generate manufacturing files, use the tool utilities to generate manufacturing GERBER files. Congratulation, you have completed your first design.

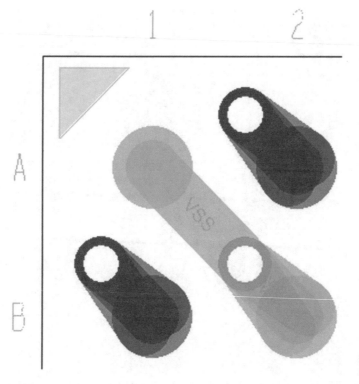

Fig. 5.24 Place a right triangle fiducial in the corner to signify package *A1* ball corner

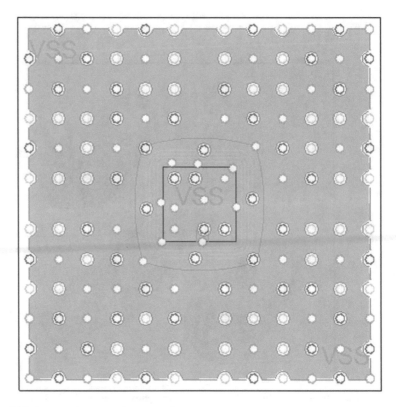

Fig. 5.25 Ground plane placed on layer-2

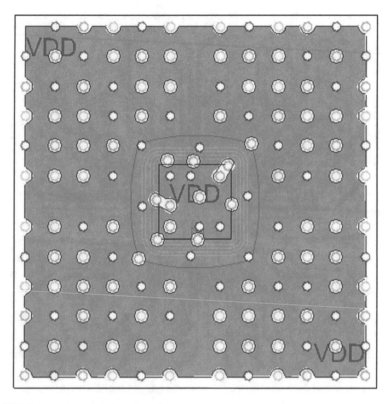

Fig. 5.26 Power plane placed on layer-3

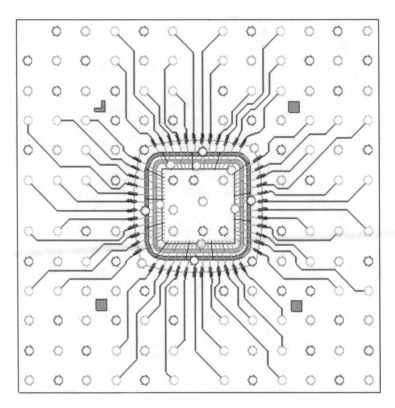

Fig. 5.27 Bondfingers are routed and connected to PTH vias on layer-1. A fully routed design

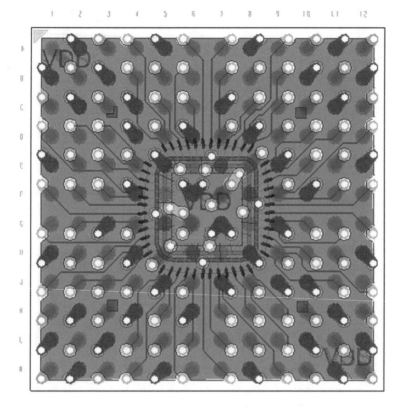

Fig. 5.28 Completed wirebond substrate design with all layers visible

5.3   Design Projects I

Exercise W1: Redesign the substrate in example-1 with BGA pitch of 1, 0.6, and 0.4 mm.

Exercise W2: Redesign the substrate in example-1 with substrate size of 8 mm × 8 mm and 6 mm × 6 mm and BGA pitch of 0.8 mm.

Exercise W3: Redesign the substrate in example-1 using blind and buried micro-vias. Can you perform the design on a smaller-size substrate such as 5 mm × 5 mm using blind and buried micro-vias and smaller BGA ball pitch such as 0.4 mm?

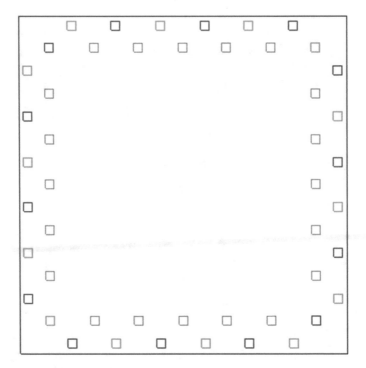

Fig. 5.A A 2 mm × 2 mm staggered pad wirebond device

Exercise W4: Redesign the substrate in example-1 with BGA signal-to-power/ground ratios of 1-to-1, 2-to-1, 3-to-1, and 4-to-1. What are the challenges in each scenario?

Exercise W5: Provide an optimal die pin assignment considering BGA ball assignment is fixed as provided in Table 5.6.

Exercise W6 (Extra Credit): Design a 10 mm × 10 mm BGA substrate with 0.6-mm ball pitch to package a 2 mm × 2 mm wirebond die with staggered bond pads (Fig. 5.A). Die pad coordinate netlist is provided in Table 5.A.

Exercise W7 (Extra Credit): Design a 15 mm × 15 mm BGA substrate with 0.6-mm ball pitch to package the staggered pad wirebond device in Exercise 6 and wirebond device presented in example-1; both dies are mounted side by side on substrate.

Exercise W8 (Extra Credit): Design a 15 mm × 15 mm BGA substrate with 0.6-mm ball pitch to package the staggered pad wirebond device in Exercise W6 mounted on substrate top layer (layer-1) with wirebond device presented in example-1, mounted on substrate bottom layer (layer-4).

Table 5.A Pad coordinate netlist for a 2 mm × 2 mm staggered pad wirebond device

Pin number	X coord	Y coord	Net name
1	−810	810	VDD
2	−675	945	VSS
3	−540	810	DIE_PIN_3
4	−405	945	VDD
5	−270	810	DIE_PIN_5
6	−135	945	VSS
7	0	810	DIE_PIN_7
8	135	945	VDD
9	270	810	DIE_PIN_9
10	405	945	VSS
11	540	810	DIE_PIN_11
12	675	945	VDD
13	810	810	VSS
14	945	675	VDD
15	810	540	DIE_PIN_15
16	945	405	VSS
17	810	270	DIE_PIN_17
18	945	135	VDD
19	810	0	DIE_PIN_19
20	945	−135	VSS
21	810	−270	DIE_PIN_21
22	945	−405	VDD
23	810	−540	DIE_PIN_23
24	945	−675	VSS
25	810	−810	VDD
26	675	−945	VSS
27	540	−810	DIE_PIN_27
28	405	−945	VDD
29	270	−810	DIE_PIN_29
30	135	−945	VSS
31	0	−810	DIE_PIN_31
32	−135	−945	VDD
33	−270	−810	DIE_PIN_33
34	−405	−945	VSS
35	−540	−810	DIE_PIN_35
36	−675	−945	VDD
37	−810	−810	VSS
38	−945	−675	VDD
39	−810	−540	DIE_PIN_39

(continued)

Table 5.A (continued)

Pin number	X coord	Y coord	Net name
40	−945	−405	VSS
41	−810	−270	DIE_PIN_41
42	−945	−135	VDD
43	−810	0	DIE_PIN_43
44	−945	135	VSS
45	−810	270	DIE_PIN_45
46	−945	405	VDD
47	−810	540	DIE_PIN_47
48	−945	675	VSS

Exercise W9 (Extra Credit): Design a 4-layer, 10 mm × 10 mm substrate with BGA pitch of 0.8 mm with a cavity to package the wirebond die in example-1. Consider the thickness of the wirebond die presented in example-1 to be 250 μm. Generate a cavity on substrate top surface, and mount the die on layer-2. What is the total height of die plus substrate with cavity compared to die and substrate without cavity?

Exercise W10 (Extra Credit): Design a 4-layer, 15 mm × 15 mm substrate with BGA pitch of 0.8 mm and cavities to package the wirebond die in example-1 and the die in Exercise W6. Consider the thickness of each die to be 200 μm. Generate a cavity on substrate top surface, and mount the wirebond die in Exercise W6 on layer-2. Generate a cavity on substrate bottom surface, and mount the wirebond die in example-1 on layer-3.

Example 2: Flip Chip Substrate Design

In this example, we will demonstrate a flip chip substrate design process in a situation where package ball-out is fixed and optimal flip chip bump pattern is desired. We will reuse the package BGA ball-out coordinate from example 1 to start with. We will place and route a 2 mm × 2 mm flip chip die on a 10 mm × 10 mm BGA substrate with 0.8-mm BGA ball pitch.

Step-1: Set up substrate cross section

Set up a 4-layer, 1 + 2 + 1 build-up substrate cross section. Similar to example-1, number of layers are determined per routing density and signal/power integrity requirement. In this example, we will use 1 + 2 + 1 build-up layers to route all the signals and supply nets. Layer stack-up is shown in Table 5.7, and signals can be routed in micro-strip format on layer-1 with reference to ground plane underneath.

   Dielectric and copper thicknesses can be varied to meet the signal impedance requirement. Similar to example-1, consider 50-Ω impedance requirement for all signals, and use an impedance calculator to determine the trace width necessary to achieve 50-Ω impedance per selected dielectric and copper thicknesses. We will use 400 μm as substrate core thickness.

Table 5.7 1 + 2 + 1 build-up substrate layer stack-up arrangement

Layer-1 (build-up layer)	Signal
Layer-2 (core layer)	Ground (vss)
Layer-3 (core layer)	Power (vdd)
Layer-4 (build-up layer)	Signal

Table 5.8 Padstacks definition for 1 + 2 + 1 flip chip build-up substrate

Padstacks	Specification (unit: μm)
Die bump landing pad	120 (circle)
Die bump pad opening	90 (circle)
BGA landing pad	450 (circle)
BGA pad opening	350 (circle)
PTH core via	200 pad, 100 drill (circle)
Micro via	85 pad, 60 drill (circle)

Step-2: Set up design rules

Similar to example-1, design rules/constraints must be step up for all layers, and this includes build-ups and core layers before proceeding with the design. As noted earlier, these rules include manufacturing, electrical, mechanical. Substrate manufacturing design rules for build-up substrates were discussed in Chap. 4. For this example, critical spacing constraints to be set up are trace-to-trace, trace-to-via, trace-to-shape, trace-to-(bump pad), via-to-via, via-to-shape, shape-to-shape, via-to-(bump pad).

Step-3: Set up padstacks

As noted in example-1, padstacks are the building blocks of any substrate, and we will be using padstacks to define, route, and design the build-up substrate. Padstacks should be created for appropriate substrate layer, and BGA landing padstack is created on build-up layer-4, whereas flip chip die padstack is created on build-up layer-1. In addition, padstacks for micro-via and core PTH via are required. For this example, create the padstacks according to Table 5.8.

Step-4: Generate BGA grid and flip chip die

We will use the BGA ball coordinate and netlist provided in Table 5.6 to generate a 10 mm × 10 mm BGA substrate footprint as illustrated in Fig. 5.29. Note that BGA padstack should be on substrate build-up layer-4. Similar to example-1, make sure fiducials for layer-1 and layer-4 are in place.

Use the flip chip die bump coordinate provided in Table 5.9 to generate a 2 mm × 2 mm flip chip die. Die bump landing padstack created in step-3 can be used to generate the flip chip die footprint (Fig. 5.30).

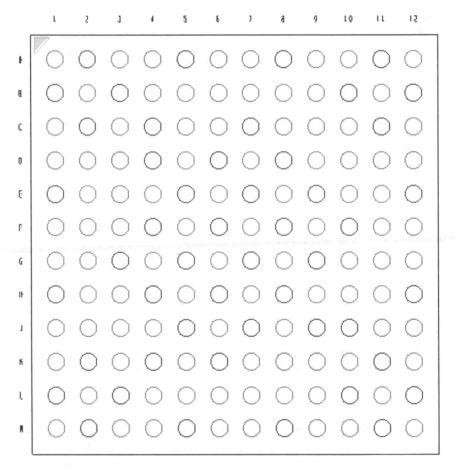

Fig. 5.29 10 mm × 10 mm substrate footprint generated according to Table 5.6

Place the flip chip die at the center of substrate, and note that flip chip die is mounted on substrate layer-1. Figure 5.15 shows the die mounted on the substrate (Fig. 5.31).

Step-6: Perform connectivity/routing

In this step, we will establish the connectivity between the BGA balls and flip chip bumps. However, first, we will investigate two scenarios to assess the effect of die bump pattern on escape routing of signals under the die. In scenario-1, signals are assigned to outer rows of the die, while power and ground are assigned to bumps at the center of the die. Assign power and ground nets to the bumps at the center of the die and 4 corners as illustrated in Fig. 5.32. In scenario-2, supply nets are assigned to the outer row bumps, while signals are assigned to bumps at the center of the die (Fig. 5.33). Assign power and ground pattern as illustrated in Fig. 5.33.

Table 5.9 2 mm × 2 mm
flip chip die bump coordinate
with 220-μm bump pitch

Pin number	X coord	Y coord
1	880	880
2	660	880
3	440	880
4	220	880
5	0	880
6	−220	880
7	−440	880
8	−660	880
9	−880	880
10	880	660
11	660	660
12	440	660
13	220	660
14	0	660
15	−220	660
16	−440	660
17	−660	660
18	−880	660
19	880	440
20	660	440
21	440	440
22	220	440
23	0	440
24	−220	440
25	−440	440
26	−660	440
27	−880	440
28	880	220
29	660	220
30	440	220
31	220	220
32	0	220
33	−220	220
34	−440	220
35	−660	220
36	−880	220

(continued)

Table 5.9 (continued)

Pin number	X coord	Y coord
37	880	0
38	660	0
39	440	0
40	220	0
41	0	0
42	−220	0
43	−440	0
44	−660	0
45	−880	0
46	880	−220
47	660	−220
48	440	−220
49	220	−220
50	0	−220
51	−220	−220
52	−440	−220
53	−660	−220
54	−880	−220
55	880	−440
56	660	−440
57	440	−440
58	220	−440
59	0	−440
60	−220	−440
61	−440	−440
62	−660	−440
63	−880	−440
64	880	−660
65	660	−660
66	440	−660
67	220	−660
68	0	−660
69	−220	−660
70	−440	−660
71	−660	−660
72	−880	−660

(continued)

Table 5.9 (continued)

Pin number	X coord	Y coord
73	880	−880
74	660	−880
75	440	−880
76	220	−880
77	0	−880
78	−220	−880
79	−440	−880
80	−660	−880
81	−880	−880

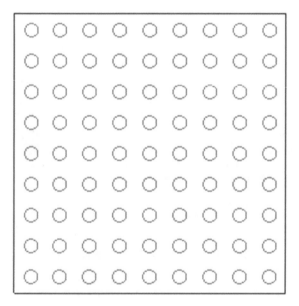

Fig. 5.30 A 2 mm × 2 mm flip chip die footprint generated based on die bump coordinate in Table 5.9

We are now in a position to route the unassigned bumps to the outer edge of the die to assess routability and congestion. Start routing the bumps on layer-1 to the die edge in any fashion, considering 18/18 μm trace width/spacing only two traces can be escaped in between the bumps. Figures 5.34 and 5.35 illustrate the escape routing under the die in scenario-1 and scenario-2, respectively. Examining both scenarios reveals that in scenario-2 almost all bump channels are filled and congested compared to scenario-1, and we will select scenario-1 bump pattern to proceed and complete our design.

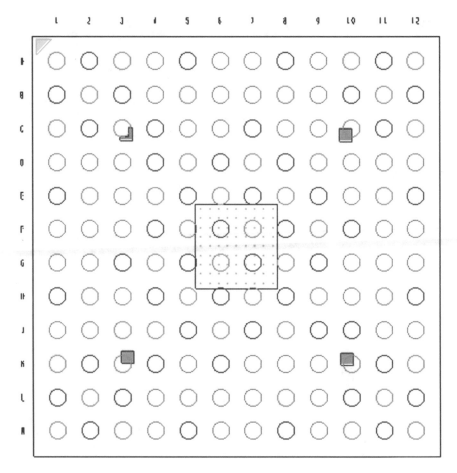

Fig. 5.31 Flip chip die positioned at the center of the substrate and mounted on build-up layer-1

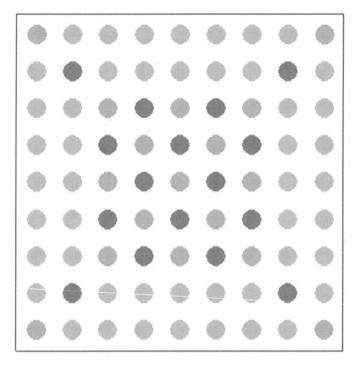

Fig. 5.32 Scenario-1 bump pattern, signals are assigned to outer rows of the die while power and ground nets are assigned to bumps at the center of the die

Similar to example-1, for each BGA signal pad drop a via from layer-4 to layer-1 and for each BGA power pad drop a via from layer-4 to layer-3; in addition, for each BGA ground pad drop a via from layer-4 to layer-2. You have effectively transitioned the power balls to layer-3, grounds balls to layer-2, and all signal balls to layer-1. Note that via in this case consists of one micro-via (layer-4 to layer-3), one PTH (layer-3 to layer-2), and one micro-via (layer-2 to layer-1), where all three vias are arranged in a staggered format to transition the signal from layer-4 to layer-1. Go ahead and assign BGA signals to die bumps, try to avoid ratnets crossover to mitigate any routing issue later on, and your design should look like Fig. 5.36. Table 5.10 provides bump coordinate netlist of the fully assigned flip chip die as depicted in Fig. 5.36. Now, complete all the signal routing by connecting all micro-vias on layer-1 to die bumps (Fig. 5.37).

We now proceed by adding the power/ground planes to the substrate. Add a power plane to layer-3 and a ground plane to layer-2. What is left is connecting the

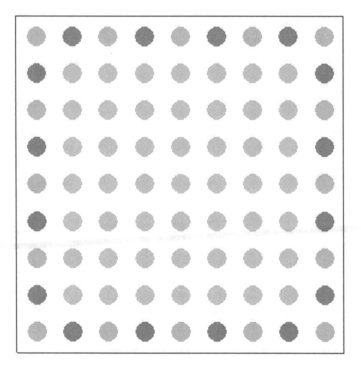

Fig. 5.33 Scenario-2 bump pattern, supply nets are assigned to the outer row bumps while signals are assigned to bumps at the center of the die

die power and ground bumps to the supply planes. Normally, one micro-via is used for every 2–3 supply bumps to route the supply bumps to supply planes. In this example, we will use one micro-via per bump and a fat trace to short the same net as illustrated in Fig. 5.38. Next, connect the power micro-vias on layer-2 to a PTH via, and this should fully connect the power bumps to the power plane on layer-3. Note that ground micro-vias are already shorted to the ground plane on layer-2. Figure 5.39 shows the ground plane, and Fig. 5.40 shows zoomed-in view of power and ground via connections on layer-2; similarly, Fig. 5.41 shows the power plane, and Fig. 5.41 shows zoomed-in view of power and ground via connections on layer-3 (Fig. 5.42).

Your design should be fully routed at this point, and Fig. 5.43 shows completed design with all layers visible.

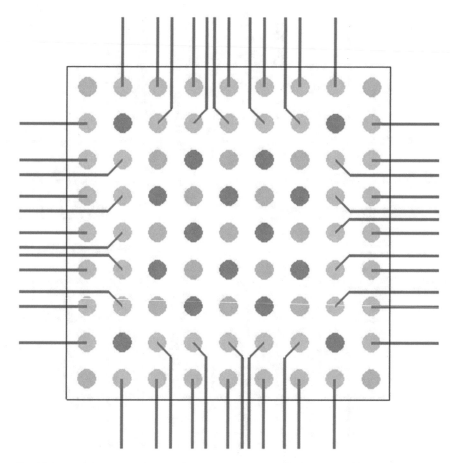

Fig. 5.34 Escape routing under the die for bump pattern presented in scenario-1 (Fig. 5.32)

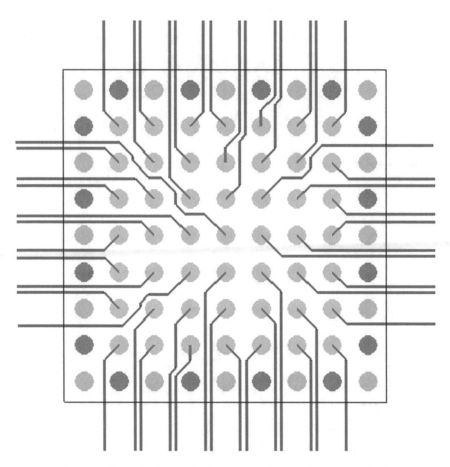

Fig. 5.35 Escape routing under the die for bump pattern presented in scenario-2 (Fig. 5.33)

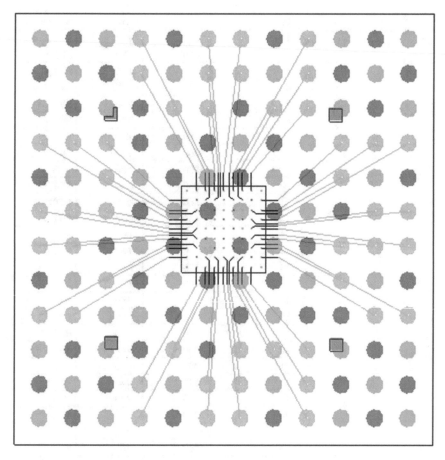

Fig. 5.36 Substrate BGA to die bump assignment with ratnets visible

Table 5.10 Fully assigned
flip chip bump coordinate
netlist as depicted in Fig. 5.36

Pin number	X coord	Y coord	Net name
1	880	880	VSS
2	660	880	SIG_1
3	440	880	SIG_4
4	220	880	SIG_7
5	0	880	SIG_10
6	−220	880	SIG_15
7	−440	880	SIG_16
8	−660	880	SIG_18
9	−880	880	VSS
10	880	660	SIG_72
11	660	660	VDD
12	440	660	SIG_3
13	220	660	SIG_6
14	0	660	SIG_9
15	−220	660	SIG_12
16	−440	660	SIG_13
17	−660	660	VDD
18	−880	660	SIG_20
19	880	440	SIG_70
20	660	440	SIG_69
21	440	440	VSS
22	220	440	VDD
23	0	440	VSS
24	−220	440	VDD
25	−440	440	VSS
26	−660	440	SIG_23
27	−880	440	SIG_21
28	880	220	SIG_67
29	660	220	SIG_66
30	440	220	VDD
31	220	220	VSS
32	0	220	VDD
33	−220	220	VSS
34	−440	220	VDD
35	−660	220	SIG_26
36	−880	220	SIG_24
37	880	0	SIG_64

(continued)

Table 5.10 (continued)

Pin number	X coord	Y coord	Net name
38	660	0	SIG_63
39	440	0	VSS
40	220	0	VDD
41	0	0	VSS
42	−220	0	VDD
43	−440	0	VSS
44	−660	0	SIG_27
45	−880	0	SIG_29
46	880	−220	SIG_61
47	660	−220	SIG_60
48	440	−220	VDD
49	220	−220	VSS
50	0	−220	VDD
51	−220	−220	VSS
52	−440	−220	VDD
53	−660	−220	SIG_32
54	−880	−220	SIG_30
55	880	−440	SIG_58
56	660	−440	SIG_57
57	440	−440	VSS
58	220	−440	VDD
59	0	−440	VSS
60	−220	−440	VDD
61	−440	−440	VSS
62	−660	−440	SIG_35
63	−880	−440	SIG_33
64	880	−660	SIG_55
65	660	−660	VDD
66	440	−660	SIG_52
67	220	−660	SIG_49
68	0	−660	SIG_44
69	−220	−660	SIG_43
70	−440	−660	SIG_40
71	−660	−660	VDD
72	−880	−660	SIG_36
73	880	−880	VSS
74	660	−880	SIG_53
75	440	−880	SIG_50
76	220	−880	SIG_47

(continued)

Table 5.10 (continued)

Pin number	X coord	Y coord	Net name
77	0	−880	SIG_46
78	−220	−880	SIG_41
79	−440	−880	SIG_38
80	−660	−880	SIG_37
81	−880	−880	VSS

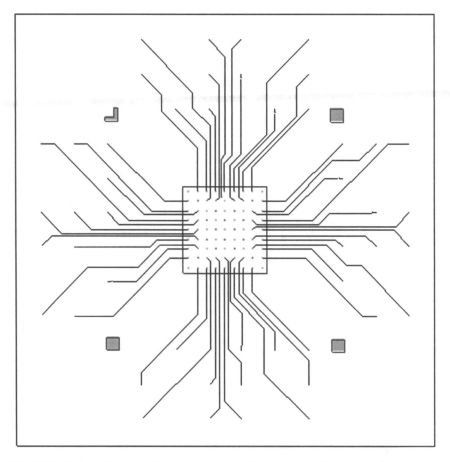

Fig. 5.37 All the device signal bumps are routed and connected to micro-vias on layer-1

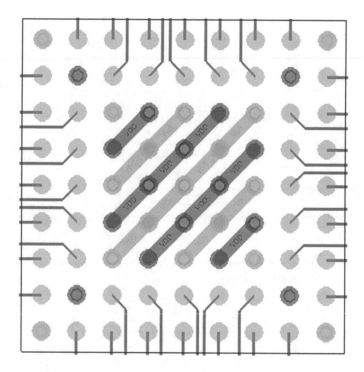

Fig. 5.38 Connecting the die power and ground bumps to the supply planes, one micro-via for every 1–2 bumps and a fat trace to short the same net

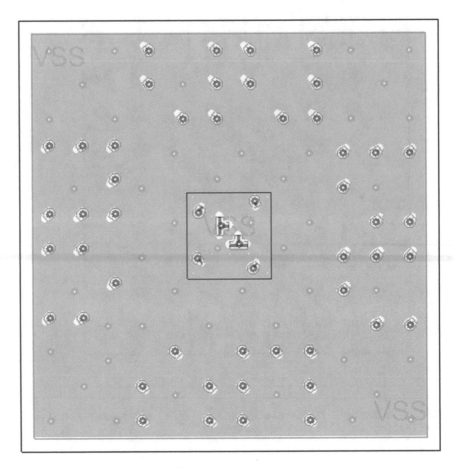

Fig. 5.39 Fully connected ground plane

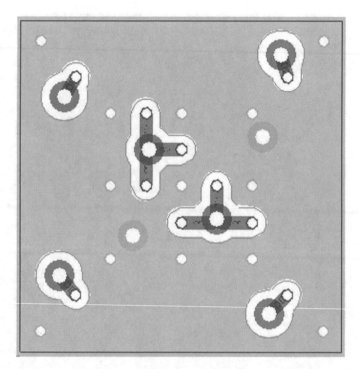

Fig. 5.40 Zoomed-in view of power and ground via connections on layer-2 right under the die

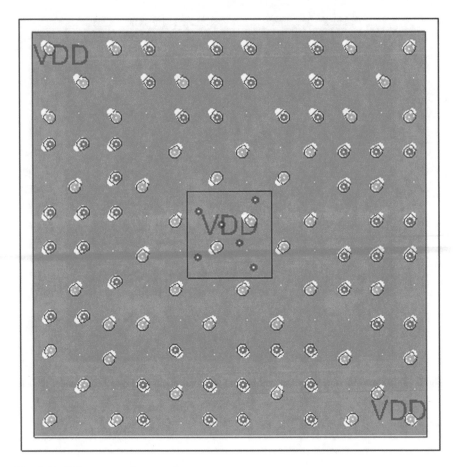

Fig. 5.41 Fully connected power plane

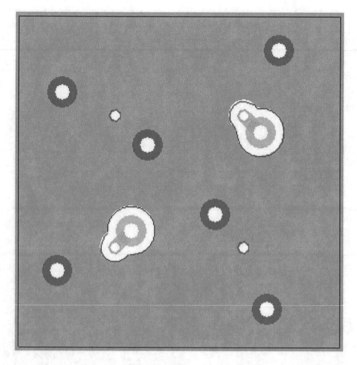

Fig. 5.42 Zoomed-in view of power and ground via connections on layer-4 right under the die

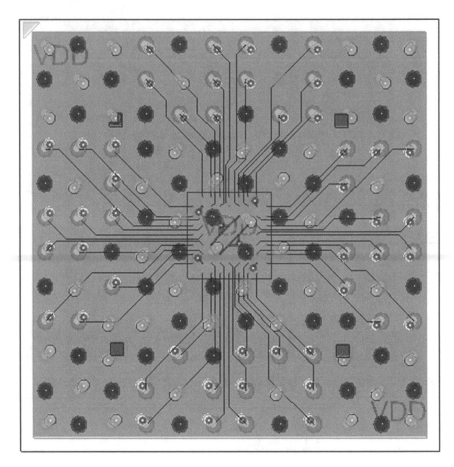

Fig. 5.43 Completed flip chip substrate design with all layers visible

Step: 7: Verification and manufacturing

We are arrived at the most important stage of the design process, the entire substrate design must be verified for any design rule violation and any signal/power integrity violation as explained in example-1. Run a comprehensive design rule check on the design, if any DRC violation is observed, identify the cause of the violation and try to resolve it. If the DRC violation is due to spacing constraint, try to slide the elements to resolve the violation. If, for example, the violation is due to timing error, vary the conductor length to resolve the timing issue. Once the design is free of any DRC violation, generate a set of report on important criteria such as conductor length, timing, and die-to-package netlist report; these reports are mandatory and required for sign-off and release the design to manufacturing. All tools provide capabilities to generate manufacturing files, use the tool utilities to generate manufacturing GERBER files. Congratulation, you have completed your first flip chip substrate design.

5.4   Design Projects II

Exercise F1: Redesign the substrate in example-2 with die bump pitch of 200, 180, 150, and 100 μm.

Exercise F2: Design a 15 mm × 15 mm BGA substrate to package dies with signal-to-power/ground ratios of 1-to-1, 2-to-1, 3-to-1, and 4-to-1. What are the challenges in each scenario?

Exercise F3: Redesign the substrate in example-2 with substrate size of 8 mm 8 mm and 6 mm × 6 mm and BGA pitch of 0.8 mm.

Exercise F4: Redesign the substrate in example-2 with substrate size of 5 mm 5 mm and BGA pitch of 0.4 mm.

Exercise F5: Redesign the substrate in example-2 with substrate size of 3 mm 3 mm and BGA pitch of 0.3 mm.

Exercise F6: Redesign the substrate in example-2 with die bump pattern as illustrated in Fig. 5.35. What are the challenges?

Exercise F7 (Extra Credit): Design a 19 mm × 19 mm substrate with BGA pitch of 0.8 mm to package both wirebond and flip chip dies presented in examples 1 and 2; dies are mounted side by side on substrate top layer (Layer-1). Make sure to rename the signal net names on one of dies; each die should have unique signal net name.

Exercise F8 (Extra Credit): Design a 19 mm × 19 mm substrate with BGA pitch of 0.8 mm to package both wirebond and flip chip dies presented in examples 1 and 2, where flip chip die is mounted on substrate layer-1 and wirebond die is mounted on the top of flip chip die (stacked configuration). Make sure to rename the signal net names on one of dies, each die should have unique signal net name.

Exercise F9 (Extra Credit): Design a 19 mm × 19 mm substrate with BGA pitch of 0.8 mm to package the staggered pad wirebond die in Exercise W6 (mounted on layer-1) and the flip chip die presented in example 2 (mounted on layer-4). Make sure to rename the signal net names on one of dies; each die should have unique signal net name.

Exercise F10 (Extra Credit): Design a 19 mm × 19 mm substrate with BGA pitch of 0.6 mm to package the staggered pad wirebond die in Exercise W6 (mounted on layer-1), flip chip die presented in example 2 (mounted on layer-1), and wirebond die presented in example-1 (mounted on layer-4). Make sure to rename the signal net name; each die should have unique signal net name.

Chapter 6
Pathfinding and Co-design

Abstract In this chapter, we will learn about die-package-board pathfinding and co-design. We will also learn about basics of device floorplanning and co-design. A number of industry-based examples are provided to demonstrate the integration process. Co-design and pathfinding methodologies presented in this chapter build the foundation of 2.5D/3D heterogeneous integration.

6.1 Co-design and Pathfinding

Pathfinding is a methodology for exploring, investigating, and planning an optimal system connectivity at the early stages of design process. Ideally, pathfinding is performed in a dynamic and fluidic type setting where designer can explorer optimal connectivity, Fig. 6.1. One of the most important roles of a package engineer is to work closely with the device team to optimize the pad/bump placement for optimal wirebonding or flip chip packaging. Device bond/bump pad placement must be optimized for wirebonding and flip chip assembly process; bond/bump pads must also be optimized for optimal signal and power integrity routing and performance of the substrate. On the other hand, package engineer interacts closely with the printed circuit board (PCB) team to ensure that package BGA ball is optimally patterned for routing on the PCB. Package engineer is expected to have good knowledge of device physical implementation as well as PCB physical implementation.

In a complex modern system design, package engineer is expected to propose an optimal device pad pattern as well as an optimal BGA ball pattern at the very early stages of the design, and this process is called Silicon-package-board co-design and pathfinding Fig. 6.2. The co-design and pathfinding processes ensure optimal end-to-end signal and power/ground connectivity across the system.

Having mastered the principals of physical design in Chap. 4, we will start the process of pathfinding and optimization with various practical industry-based examples.

© Springer International Publishing AG 2018
F. Yazdani, *Foundations of Heterogeneous Integration: An Industry-Based, 2.5D/3D Pathfinding and Co-Design Approach*,
https://doi.org/10.1007/978-3-319-75769-8_6

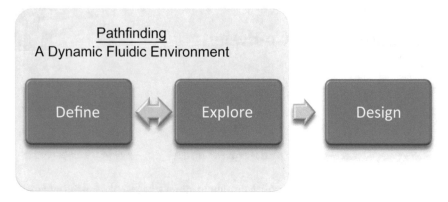

Fig. 6.1 Role of pathfinding in the design process

Fig. 6.2 Role of package design team in silicon-package-board co-design and pathfinding

6.2   Example-1: An MCM Pathfinding/Optimization

In this example, we will define and perform pathfinding to optimize a Multi-Chip-Module (MCM) package. Consider the scenario where we have to design a package to accommodate two flip chip devices mounted on a BGA package, where the package is interfaced with a connector on PCB. Connector is an off-the-shelf component with fixed footprint, Fig. 6.3 and fixed netlist, Table 6.1. One of the devices (Die-2) has already been manufactured with fixed bump pattern/ netlist Fig. 6.4 and Table 6.2, whereas another flip chip device (Die-1) is at the early stages of design process, Fig. 6.5 and Table 6.3.

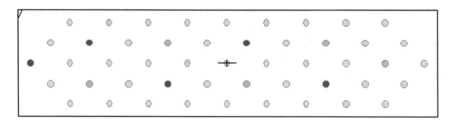

Fig. 6.3 Off-the-shelf connector with fixed footprint and netlist

Pin number	x-coor	y-coor	Net name
			Table 6.1 Connector predefined coordinate netlist
1	−703.96	1070.03	SIG_B[39]
2	−503.96	1070.03	SIG_B[38]
3	−303.96	1070.03	SIG_B[37]
4	−103.96	1070.03	SIG_B[36]
5	96.04	1070.03	SIG_B[35]
6	296.04	1070.03	SIG_B[34]
7	496.04	1070.03	SIG_B[33]
8	696.04	1070.03	SIG_B[32]
9	896.04	1070.03	SIG_B[31]
10	−803.96	970.02	SIG_B[30]
11	−603.96	970.02	VDD
12	−403.96	970.02	SIG_B[29]
13	−203.96	970.02	VSS
14	−3.96	970.02	SIG_B[28]
15	196.04	970.02	VDD
16	396.04	970.02	SIG_B[27]
17	596.04	970.02	VSS
18	796.04	970.02	SIG_B[26]
19	996.04	970.02	SIG_B[25]
20	−903.96	870.02	VDD
21	−703.96	870.02	SIG_B[24]
22	−503.96	870.02	SIG_B[23]
23	−303.96	870.02	SIG_B[22]
24	−103.96	870.02	SIG_B[21]
25	96.04	870.02	SIG_B[20]
26	296.04	870.02	SIG_B[19]
27	496.04	870.02	SIG_B[18]
28	696.04	870.02	SIG_B[17]
29	896.04	870.02	VSS
30	1096.04	870.02	SIG_B[16]
31	−803.96	770.02	SIG_B[15]
32	−603.96	770.02	VSS
33	−403.96	770.02	SIG_B[14]
34	−203.96	770.02	VDD
35	−3.96	770.02	SIG_B[13]
36	196.04	770.02	VSS
37	396.04	770.02	SIG_B[12]
38	596.04	770.02	VDD
39	796.04	770.02	SIG_B[11]
40	996.04	770.02	SIG_B[10]

(continued)

Table 6.1 (continued)

Pin number	x-coor	y-coor	Net name
41	−703.96	670.02	SIG_B[9]
42	−503.96	670.02	SIG_B[8]
43	−303.96	670.02	SIG_B[7]
44	−103.96	670.02	SIG_B[6]
45	96.04	670.02	SIG_B[5]
46	296.04	670.02	SIG_B[4]
47	496.04	670.02	SIG_B[3]
48	696.04	670.02	SIG_B[2]
49	896.04	670.02	SIG_B[1]

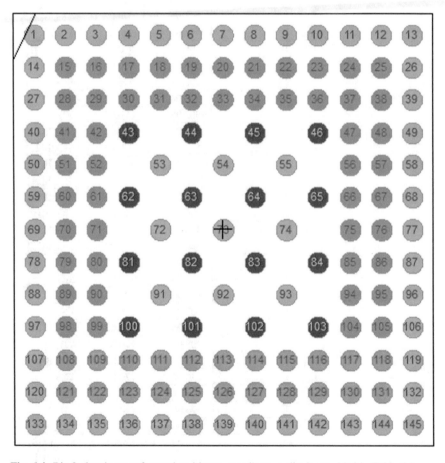

Fig. 6.4 Die-2 already manufactured and bump coordinate netlist is provided in Table 6.1

Table 6.2 Die-2 bump coordinate netlist as illustrated in Fig. 6.3

Pin number	x-coor	y-coor	Net name
1	−900	900	VSS
2	−750	900	VSS
3	−600	900	VSS
4	−450	900	VSS
5	−300	900	VSS
6	−150	900	VSS
7	0	900	VSS
8	150	900	VSS
9	300	900	VSS
10	450	900	VSS
11	600	900	VSS
12	750	900	VSS
13	900	900	VSS
14	−900	750	VSS
15	−750	750	ddr[1]
16	−600	750	ddr[2]
17	−450	750	ddr[3]
18	−300	750	ddr[4]
19	−150	750	ddr[5]
20	0	750	ddr[6]
21	150	750	ddr[7]
22	300	750	ddr[8]
23	450	750	ddr[9]
24	600	750	ddr[10]
25	750	750	ddr[11]
26	900	750	VSS
27	−900	600	VSS
28	−750	600	ddr[12]
29	−600	600	ddr[13]
30	−450	600	ddr[14]
31	−300	600	ddr[15]
32	−150	600	ddr[16]
33	0	600	ddr[17]
34	150	600	ddr[18]
35	300	600	ddr[19]
36	450	600	ddr[20]
37	600	600	ddr[21]
38	750	600	ddr[22]
39	900	600	VSS
40	−900	450	VSS

(continued)

Table 6.2 (continued)

Pin number	x-coor	y-coor	Net name
41	−750	450	ddr[23]
42	−600	450	ddr[24]
43	−450	450	VDD
44	−150	450	VDD
45	150	450	VDD
46	450	450	VDD
47	600	450	ddr[25]
48	750	450	ddr[26]
49	900	450	VSS
50	−900	300	VSS
51	−750	300	ddr[27]
52	−600	300	ddr[28]
53	−300	300	VSS
54	0	300	VSS
55	300	300	VSS
56	600	300	ddr[29]
57	750	300	ddr[30]
58	900	300	VSS
59	−900	150	VSS
60	−750	150	ddr[31]
61	−600	150	ddr[32]
62	−450	150	VDD
63	−150	150	VDD
64	150	150	VDD
65	450	150	VDD
66	600	150	ddr[33]
67	750	150	ddr[34]
68	900	150	VSS
69	−900	0	VSS
70	−750	0	ddr[35]
71	−600	0	ddr[36]
72	−300	0	VSS
73	0	0	VSS
74	300	0	VSS
75	600	0	ddr[37]
76	750	0	ddr[38]
77	900	0	VSS
78	−900	−150	VSS
79	−750	−150	ddr[39]
80	−600	−150	ddr[40]

(continued)

Table 6.2 (continued)

Pin number	x-coor	y-coor	Net name
81	−450	−150	VDD
82	−150	−150	VDD
83	150	−150	VDD
84	450	−150	VDD
85	600	−150	ddr[41]
86	750	−150	ddr[42]
87	900	−150	VSS
88	−900	−300	VSS
89	−750	−300	ddr[43]
90	−600	−300	ddr[44]
91	−300	−300	VSS
92	0	−300	VSS
93	300	−300	VSS
94	600	−300	ddr[45]
95	750	−300	ddr[46]
96	900	−300	VSS
97	−900	−450	VSS
98	−750	−450	ddr[47]
99	−600	−450	ddr[48]
100	−450	−450	VDD
101	−150	−450	VDD
102	150	−450	VDD
103	450	−450	VDD
104	600	−450	ddr[49]
105	750	−450	ddr[50]
106	900	−450	VSS
107	−900	−600	VSS
108	−750	−600	add[1]
109	−600	−600	add[2]
110	−450	−600	add[3]
111	−300	−600	add[4]
112	−150	−600	add[5]
113	0	−600	add[6]
114	150	−600	add[7]
115	300	−600	add[8]
116	450	−600	add[9]
117	600	−600	add[10]
118	750	−600	add[11]
119	900	−600	VSS
120	−900	−750	VSS

(continued)

Table 6.2 (continued)

Pin number	x-coor	y-coor	Net name
121	−750	−750	add[12]
122	−600	−750	add[13]
123	−450	−750	add[14]
124	−300	−750	add[15]
125	−150	−750	add[16]
126	0	−750	add[17]
127	150	−750	add[18]
128	300	−750	add[19]
129	450	−750	add[20]
130	600	−750	add[21]
131	750	−750	add[22]
132	900	−750	VSS
133	−900	−900	VSS
134	−750	−900	VSS
135	−600	−900	VSS
136	−450	−900	VSS
137	−300	−900	VSS
138	−150	−900	VSS
139	0	−900	VSS
140	150	−900	VSS
141	300	−900	VSS
142	450	−900	VSS
143	600	−900	VSS
144	750	−900	VSS
145	900	−900	VSS

The package design team is called upon to propose a bump pattern/netlist for Die-1; the package size and BGA ball pitch are defined by the customer and are fixed; however, BGA ball netlist is not defined and is up to the package design team to propose one. In summary, package design team must propose an optimal bump connectivity netlist for Die-1 and complete package BGA ball-out. Both devices are 2 mm × 2 mm in size with 150 μm bump pitch, and the package body size is defined to be 12 mm × 12 mm with 0.8 mm BGA ball pitch.

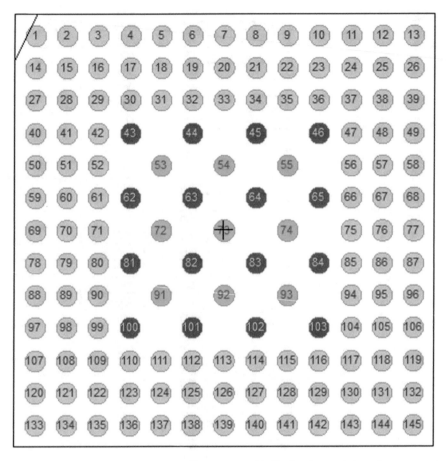

Fig. 6.5 Die-1 is at the early stage of design process with die size, bump pitch, and core pattern/pitch already defined

Furthermore, Die-2 device team has already communicated to the package team that the parallel bus interface (add[ ] net) on the device should route and connect directly to Die-1, meaning these signals do not connect to any BGA balls. The rest of the signals (ddr[ ] net) on Device-2 should be routed and connected to package BGA balls.

Table 6.3 Die-1 bump coordinate netlist at the early stages of design

Pin number	x-coor	y-coor	Net name
1	−900	900	
2	−750	900	
3	−600	900	
4	−450	900	
5	−300	900	
6	−150	900	
7	0	900	
8	150	900	
9	300	900	
10	450	900	
11	600	900	
12	750	900	
13	900	900	
14	−900	750	
15	−750	750	
16	−600	750	
17	−450	750	
18	−300	750	
19	−150	750	
20	0	750	
21	150	750	
22	300	750	
23	450	750	
24	600	750	
25	750	750	
26	900	750	
27	−900	600	
28	−750	600	
29	−600	600	
30	−450	600	
31	−300	600	
32	−150	600	
33	0	600	
34	150	600	
35	300	600	
36	450	600	
37	600	600	
38	750	600	
39	900	600	
40	−900	450	

(continued)

Table 6.3 (continued)

Pin number	x-coor	y-coor	Net name
41	−750	450	
42	−600	450	
43	−450	450	VDD
44	−150	450	VDD
45	150	450	VDD
46	450	450	VDD
47	600	450	
48	750	450	
49	900	450	
50	−900	300	
51	−750	300	
52	−600	300	
53	−300	300	VSS
54	0	300	VSS
55	300	300	VSS
56	600	300	
57	750	300	
58	900	300	
59	−900	150	
60	−750	150	
61	−600	150	
62	−450	150	VDD
63	−150	150	VDD
64	150	150	VDD
65	450	150	VDD
66	600	150	
67	750	150	
68	900	150	
69	−900	0	
70	−750	0	
71	−600	0	
72	−300	0	VSS
73	0	0	VSS
74	300	0	VSS
75	600	0	
76	750	0	
77	900	0	
78	−900	−150	
79	−750	−150	
80	−600	−150	

(continued)

Table 6.3 (continued)

Pin number	x-coor	y-coor	Net name
81	−450	−150	VDD
82	−150	−150	VDD
83	150	−150	VDD
84	450	−150	VDD
85	600	−150	
86	750	−150	
87	900	−150	
88	−900	−300	
89	−750	−300	
90	−600	−300	
91	−300	−300	VSS
92	0	−300	VSS
93	300	−300	VSS
94	600	−300	
95	750	−300	
96	900	−300	
97	−900	−450	
98	−750	−450	
99	−600	−450	
100	−450	−450	VDD
101	−150	−450	VDD
102	150	−450	VDD
103	450	−450	VDD
104	600	−450	
105	750	−450	
106	900	−450	
107	−900	−600	
108	−750	−600	
109	−600	−600	
110	−450	−600	
111	−300	−600	
112	−150	−600	
113	0	−600	
114	150	−600	
115	300	−600	
116	450	−600	
117	600	−600	
118	750	−600	
119	900	−600	
120	−900	−750	

(continued)

Table 6.3 (continued)

Pin number	x-coor	y-coor	Net name
121	−750	−750	
122	−600	−750	
123	−450	−750	
124	−300	−750	
125	−150	−750	
126	0	−750	
127	150	−750	
128	300	−750	
129	450	−750	
130	600	−750	
131	750	−750	
132	900	−750	
133	−900	−900	
134	−750	−900	
135	−600	−900	
136	−450	−900	
137	−300	−900	
138	−150	−900	
139	0	−900	
140	150	−900	
141	300	−900	
142	450	−900	
143	600	−900	
144	750	−900	
145	900	−900	

Guidelines from Die-1 design team state that there are four signals groups on the device, namely Signal group add[ ], Signal group A (SIG_A), Signal group B (SIG_B), and Signal group C (SIG_C). SIG_A and SIG_C group should be routed and connected to the package BGA balls, whereas SIG_B group is routed to package BGA balls and routed to the connector on the PCB, Fig. 6.3.

We will first generate a 12 mm × 12 mm substrate with 0.8 mm BGA ball pitch, and detailed substrate cross section and design rules can be defined later during the design process, but at this stage, we would like to establish an optimal bump pattern connectivity among the components. Assign power and ground nets

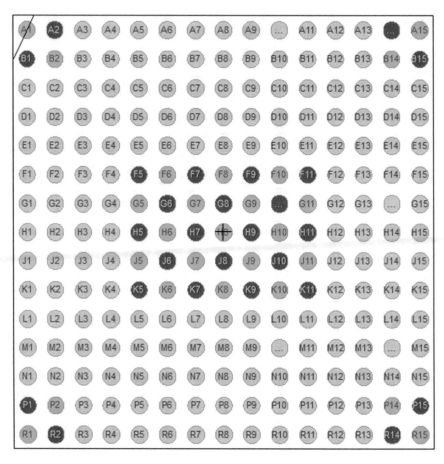

Fig. 6.6 12 mm × 12 mm substrate with 0.8 mm BGA ball pitch. Supply nets are assigned at the center and four corners

to BGA balls on four corners of the substrate as well as at the center of the substrate right under the die as illustrated in Fig. 6.6; BGA netlist is provided in Table 6.4. Generate devices and connector using the coordinate data are provided in Tables 6.1, 6.2, and 6.3. Mount both devices at 1 mm apart on substrate top layer positioned at the center of the substrate, and rotate Die-2 such that add[ ] interface faces Die-1. Mount the connector on substrate bottom layer and some distance apart from the substrate; the arrangement should look like Fig. 6.7.

Table 6.4 BGA coordinate data for 12 mm × 12 mm substrate with 0.8 mm BGA ball pitch depicted in Fig. 6.6

Pin number	x-coor	y-coor	Net name
A1	−5600	5600	VSS
A2	−4800	5600	VDD
A3	−4000	5600	
A4	−3200	5600	
A5	−2400	5600	
A6	−1600	5600	
A7	−800	5600	
A8	0	5600	
A9	800	5600	
A10	1600	5600	
A11	2400	5600	
A12	3200	5600	
A13	4000	5600	
A14	4800	5600	VDD
A15	5600	5600	VSS
B1	−5600	4800	VDD
B2	−4800	4800	VSS
B3	−4000	4800	
B4	−3200	4800	
B5	−2400	4800	
B6	−1600	4800	
B7	−800	4800	
B8	0	4800	
B9	800	4800	
B10	1600	4800	
B11	2400	4800	
B12	3200	4800	
B13	4000	4800	
B14	4800	4800	VSS
B15	5600	4800	VDD
C1	−5600	4000	
C2	−4800	4000	
C3	−4000	4000	
C4	−3200	4000	
C5	−2400	4000	
C6	−1600	4000	
C7	−800	4000	
C8	0	4000	
C9	800	4000	
C10	1600	4000	
C11	2400	4000	
C12	3200	4000	

(continued)

Table 6.4 (continued)

Pin number	x-coor	y-coor	Net name
C13	4000	4000	
C14	4800	4000	
C15	5600	4000	
D1	−5600	3200	
D2	−4800	3200	
D3	−4000	3200	
D4	−3200	3200	
D5	−2400	3200	
D6	−1600	3200	
D7	−800	3200	
D8	0	3200	
D9	800	3200	
D10	1600	3200	
D11	2400	3200	
D12	3200	3200	
D13	4000	3200	
D14	4800	3200	
D15	5600	3200	
E1	−5600	2400	
E2	−4800	2400	
E3	−4000	2400	
E4	−3200	2400	
E5	−2400	2400	
E6	−1600	2400	
E7	−800	2400	
E8	0	2400	
E9	800	2400	
E10	1600	2400	
E11	2400	2400	
E12	3200	2400	
E13	4000	2400	
E14	4800	2400	
E15	5600	2400	
F1	−5600	1600	
F2	−4800	1600	
F3	−4000	1600	
F4	−3200	1600	
F5	−2400	1600	VDD
F6	−1600	1600	VSS
F7	−800	1600	VDD
F8	0	1600	VSS
F9	800	1600	VDD

(continued)

Table 6.4 (continued)

Pin number	x-coor	y-coor	Net name
F10	1600	1600	VSS
F11	2400	1600	VDD
F12	3200	1600	
F13	4000	1600	
F14	4800	1600	
F15	5600	1600	
G1	−5600	800	
G2	−4800	800	
G3	−4000	800	
G4	−3200	800	
G5	−2400	800	VSS
G6	−1600	800	VDD
G7	−800	800	VSS
G8	0	800	VDD
G9	800	800	VSS
G10	1600	800	VDD
G11	2400	800	VSS
G12	3200	800	
G13	4000	800	
G14	4800	800	
G15	5600	800	
H1	−5600	0	
H2	−4800	0	
H3	−4000	0	
H4	−3200	0	
H5	−2400	0	VDD
H6	−1600	0	VSS
H7	−800	0	VDD
H8	0	0	VSS
H9	800	0	VDD
H10	1600	0	VSS
H11	2400	0	VDD
H12	3200	0	
H13	4000	0	
H14	4800	0	
H15	5600	0	
J1	−5600	−800	
J2	−4800	−800	
J3	−4000	−800	
J4	−3200	−800	
J5	−2400	−800	VSS

(continued)

Table 6.4 (continued)

Pin number	x-coor	y-coor	Net name
J6	−1600	−800	VDD
J7	−800	−800	VSS
J8	0	−800	VDD
J9	800	−800	VSS
J10	1600	−800	VDD
J11	2400	−800	VSS
J12	3200	−800	
J13	4000	−800	
J14	4800	−800	
J15	5600	−800	
K1	−5600	−1600	
K2	−4800	−1600	
K3	−4000	−1600	
K4	−3200	−1600	
K5	−2400	−1600	VDD
K6	−1600	−1600	VSS
K7	−800	−1600	VDD
K8	0	−1600	VSS
K9	800	−1600	VDD
K10	1600	−1600	VSS
K11	2400	−1600	VDD
K12	3200	−1600	
K13	4000	−1600	
K14	4800	−1600	
K15	5600	−1600	
L1	−5600	−2400	
L2	−4800	−2400	
L3	−4000	−2400	
L4	−3200	−2400	
L5	−2400	−2400	
L6	−1600	−2400	
L7	−800	−2400	
L8	0	−2400	
L9	800	−2400	
L10	1600	−2400	
L11	2400	−2400	
L12	3200	−2400	
L13	4000	−2400	
L14	4800	−2400	
L15	5600	−2400	
M1	−5600	−3200	

(continued)

Table 6.4 (continued)

Pin number	x-coor	y-coor	Net name
M2	−4800	−3200	
M3	−4000	−3200	
M4	−3200	−3200	
M5	−2400	−3200	
M6	−1600	−3200	
M7	−800	−3200	
M8	0	−3200	
M9	800	−3200	
M10	1600	−3200	
M11	2400	−3200	
M12	3200	−3200	
M13	4000	−3200	
M14	4800	−3200	
M15	5600	−3200	
N1	−5600	−4000	
N2	−4800	−4000	
N3	−4000	−4000	
N4	−3200	−4000	
N5	−2400	−4000	
N6	−1600	−4000	
N7	−800	−4000	
N8	0	−4000	
N9	800	−4000	
N10	1600	−4000	
N11	2400	−4000	
N12	3200	−4000	
N13	4000	−4000	
N14	4800	−4000	
N15	5600	−4000	
P1	−5600	−4800	VDD
P2	−4800	−4800	VSS
P3	−4000	−4800	
P4	−3200	−4800	
P5	−2400	−4800	
P6	−1600	−4800	
P7	−800	−4800	
P8	0	−4800	
P9	800	−4800	
P10	1600	−4800	
P11	2400	−4800	
P12	3200	−4800	

(continued)

Table 6.4 (continued)

Pin number	x-coor	y-coor	Net name
P13	4000	−4800	
P14	4800	−4800	VSS
P15	5600	−4800	VDD
R1	−5600	−5600	VSS
R2	−4800	−5600	VDD
R3	−4000	−5600	
R4	−3200	−5600	
R5	−2400	−5600	
R6	−1600	−5600	
R7	−800	−5600	
R8	0	−5600	
R9	800	−5600	
R10	1600	−5600	
R11	2400	−5600	
R12	3200	−5600	
R13	4000	−5600	
R14	4800	−5600	VDD
R15	5600	−5600	VSS

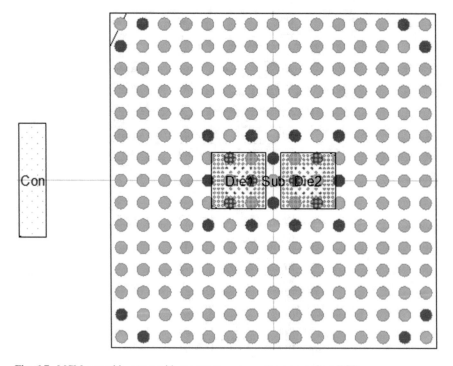

Fig. 6.7 MCM assembly setup with respect to a connector mounted on PCB

We are now at the point to assign signals to Die-1 and BGA balls while considering routability to Die-1 and connector. In order to accommodate the four signal groups, we will segment Die-1 into four banks; Bank-1 would be identical to add[ ] bus interface in Die-2; Bank-2 and Bank-3 would accommodate signal groups SIG [A] and SIG[C] while Bank-4 houses SIG[B]. Since at this stage we really do not know which signal should be assigned to which bump, thus, we will randomly assign the signal groups to bumps within each Bank; Die-1 should look like Fig. 6.8. Note that signal group add[ ] is shielded with ground bumps similar to same pattern as Die-2, and this means these signals will be routed on layer-2 as stripline with reference to ground plan on layer-1.

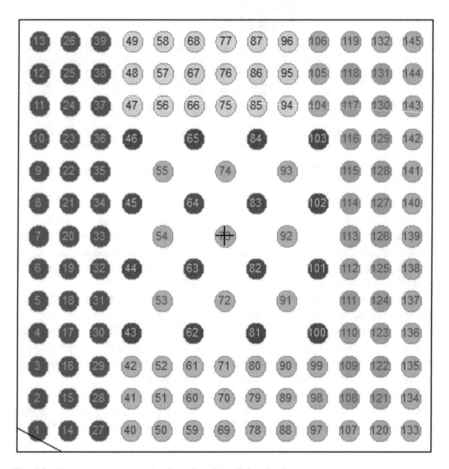

Fig. 6.8 Signal groups are randomly assigned to all four banks

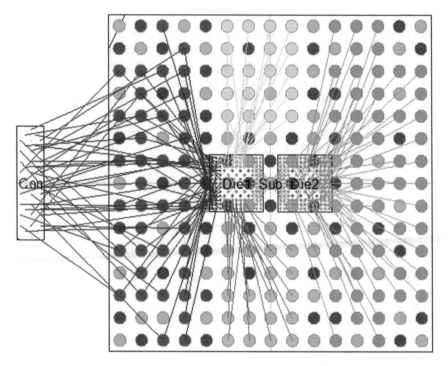

Fig. 6.9 Ratnets (flightlines) depicting unoptimized system as a result of random signal assignment during pathfinding

We will now work on the BGA pattern by assigning power and grounds to BGA foot print, and you can try different signal to power/ground ratio and study the effect. Obviously, adequate supply balls are needed to insure clean power delivery to devices. Next, we will randomly assign all the signals from Die-1 and Die-2 to BGA balls; Fig. 6.9 illustrates the ratnets (flightlines) at this stage.

As noted earlier, components with fix pattern/netlist drive the connectivity; in this system, components with fix pattern/netlist are Die-2 and connector, thus, we will optimize the BGA balls and Die-1 bumps with respect to Die-2 and connector. The tangled ratnets as depicted in Fig. 6.9 suggest serious difficulty during substrate routing. We will optimize the connectivity by swapping the BGA balls and Die-1 bumps until we achieve fully untangled ratnets (flightlines).

We will start the optimization process by swapping the BGA balls with respect to Die-2, ddr[ ] signals, and we will continue swapping until ratnets are fully untangled or very minimally tangled, Fig. 6.10. Similarly, we will swap the BGA

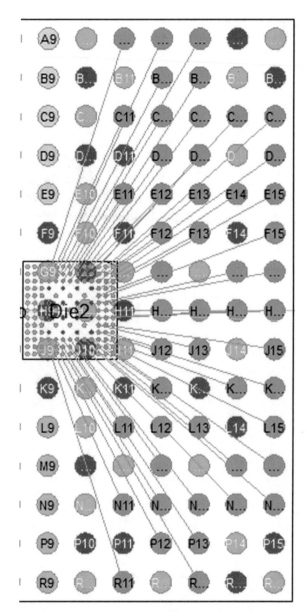

Fig. 6.10 Swapping BGA balls with respect to Die-2 ddr [ ] signals

Fig. 6.11 Swapping BGA
balls with respect to connector

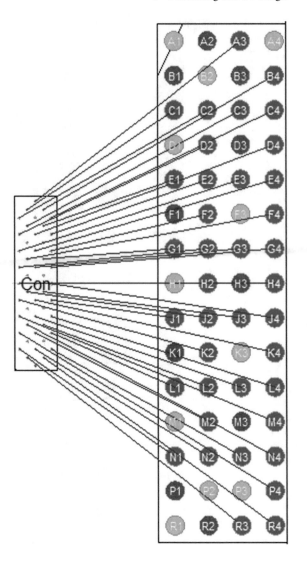

balls with respect to the connector, Fig. 6.11. *Having optimized the BGA balls with respect to fix components, we consider BGA pattern as fixed at this point and focus on swapping and optimizing Die-1 bumps with respect to BGA pattern; in addition, we will swap and optimize Die-1 add[ ] interface bumps with respect to Die-2,* Fig. 6.12. *Fully optimized Die-1 bump coordinate netlist is provided in* Table 6.5. *The fully optimized MCM package is illustrated in* Fig. 6.13, *and BGA coordinate netlist is provided in* Table 6.6.

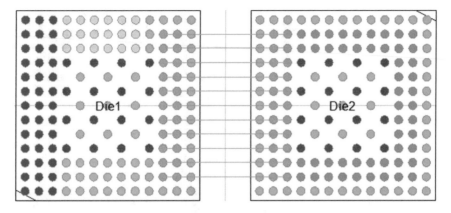

Fig. 6.12 Swapping Die-1 add[ ] interface bumps with respect to Die-2

Table 6.5 Fully optimized Die-1 bump coordinate netlist

Pin number	x-coor	y-coor	Net name
1	−900	900	SIG_B[7]
2	−750	900	SIG_B[39]
3	−600	900	SIG_B[23]
4	−450	900	SIG_B[14]
5	−300	900	SIG_B[36]
6	−150	900	SIG_B[28]
7	0	900	SIG_B[20]
8	150	900	SIG_B[12]
9	300	900	SIG_B[4]
10	450	900	SIG_B[11]
11	600	900	SIG_B[33]
12	750	900	SIG_B[25]
13	900	900	SIG_B[18]
14	−900	750	SIG_B[30]
15	−750	750	SIG_B[37]
16	−600	750	SIG_B[24]
17	−450	750	SIG_B[22]
18	−300	750	SIG_B[8]
19	−150	750	SIG_B[13]
20	0	750	SIG_B[35]
21	150	750	SIG_B[34]
22	300	750	SIG_B[3]
23	450	750	SIG_B[27]
24	600	750	SIG_B[31]
25	750	750	SIG_B[26]

(continued)

Table 6.5 (continued)

Pin number	x-coor	y-coor	Net name
26	900	750	SIG_B[17]
27	−900	600	SIG_B[15]
28	−750	600	SIG_B[38]
29	−600	600	SIG_B[29]
30	−450	600	SIG_B[9]
31	−300	600	SIG_B[21]
32	−150	600	SIG_B[5]
33	0	600	SIG_B[6]
34	150	600	SIG_B[19]
35	300	600	SIG_B[2]
36	450	600	SIG_B[32]
37	600	600	SIG_B[16]
38	750	600	SIG_B[1]
39	900	600	SIG_B[10]
40	−900	450	SIG_C[13]
41	−750	450	SIG_C[9]
42	−600	450	SIG_C[6]
43	−450	450	VDD
44	−150	450	VDD
45	150	450	VDD
46	450	450	VDD
47	600	450	SIG_A[29]
48	750	450	SIG_A[38]
49	900	450	SIG_A[45]
50	−900	300	SIG_C[12]
51	−750	300	SIG_C[2]
52	−600	300	SIG_C[17]
53	−300	300	VSS
54	0	300	VSS
55	300	300	VSS
56	600	300	SIG_A[8]
57	750	300	SIG_A[12]
58	900	300	SIG_A[46]
59	−900	150	SIG_C[8]
60	−750	150	SIG_C[5]
61	−600	150	SIG_C[11]
62	−450	150	VDD
63	−150	150	VDD
64	150	150	VDD
65	450	150	VDD

(continued)

Table 6.5 (continued)

Pin number	x-coor	y-coor	Net name
66	600	150	SIG_A[49]
67	750	150	SIG_A[14]
68	900	150	SIG_A[33]
69	−900	0	SIG_C[16]
70	−750	0	SIG_C[4]
71	−600	0	SIG_C[10]
72	−300	0	VSS
73	0	0	VSS
74	300	0	VSS
75	600	0	SIG_A[34]
76	750	0	SIG_A[41]
77	900	0	SIG_A[9]
78	−900	−150	SIG_C[1]
79	−750	−150	SIG_C[7]
80	−600	−150	SIG_C[15]
81	−450	−150	VDD
82	−150	−150	VDD
83	150	−150	VDD
84	450	−150	VDD
85	600	−150	SIG_A[10]
86	750	−150	SIG_A[42]
87	900	−150	SIG_A[50]
88	−900	−300	SIG_C[3]
89	−750	−300	SIG_C[0]
90	−600	−300	SIG_C[14]
91	−300	−300	VSS
92	0	−300	VSS
93	300	−300	VSS
94	600	−300	SIG_A[11]
95	750	−300	SIG_A[25]
96	900	−300	SIG_A[37]
97	−900	−450	VSS
98	−750	−450	VSS
99	−600	−450	VSS
100	−450	−450	VDD
101	−150	−450	VDD
102	150	−450	VDD
103	450	−450	VDD
104	600	−450	VSS
105	750	−450	VSS

(continued)

Table 6.5 (continued)

Pin number	x-coor	y-coor	Net name
106	900	−450	VSS
107	−900	−600	VSS
108	−750	−600	add[11]
109	−600	−600	add[10]
110	−450	−600	add[9]
111	−300	−600	add[8]
112	−150	−600	add[7]
113	0	−600	add[6]
114	150	−600	add[5]
115	300	−600	add[4]
116	450	−600	add[3]
117	600	−600	add[2]
118	750	−600	add[1]
119	900	−600	VSS
120	−900	−750	VSS
121	−750	−750	add[22]
122	−600	−750	add[21]
123	−450	−750	add[20]
124	−300	−750	add[19]
125	−150	−750	add[18]
126	0	−750	add[17]
127	150	−750	add[16]
128	300	−750	add[15]
129	450	−750	add[14]
130	600	−750	add[13]
131	750	−750	add[12]
132	900	−750	VSS
133	−900	−900	VSS
134	−750	−900	VSS
135	−600	−900	VSS
136	−450	−900	VSS
137	−300	−900	VSS
138	−150	−900	VSS
139	0	−900	VSS
140	150	−900	VSS
141	300	−900	VSS
142	450	−900	VSS
143	600	−900	VSS
144	750	−900	VSS
145	900	−900	VSS

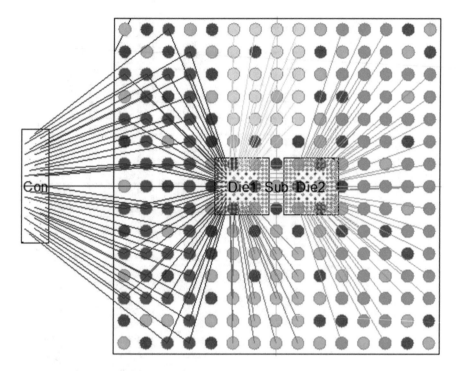

Fig. 6.13 Fully optimized MCM package

Table 6.6 Fully optimized MCM package BGA netlist

Pin number	x-coor	y-coor	Net name
A1	−5600	5600	VSS
A2	−4800	5600	VDD
A3	−4000	5600	SIG_B[16]
A4	−3200	5600	VSS
A5	−2400	5600	VDD
A6	−1600	5600	SIG_A[45]
A7	−800	5600	SIG_A[46]
A8	0	5600	SIG_A[49]
A9	800	5600	SIG_A[50]
A10	1600	5600	VSS
A11	2400	5600	ddr[47]
A12	3200	5600	ddr[43]
A13	4000	5600	ddr[39]
A14	4800	5600	VDD
A15	5600	5600	VSS
B1	−5600	4800	VDD
B2	−4800	4800	VSS

(continued)

Table 6.6 (continued)

Pin number	x-coor	y-coor	Net name
B3	−4000	4800	VDD
B4	−3200	4800	SIG_B[10]
B5	−2400	4800	VSS
B6	−1600	4800	SIG_A[38]
B7	−800	4800	VDD
B8	0	4800	SIG_A[41]
B9	800	4800	SIG_A[42]
B10	1600	4800	VDD
B11	2400	4800	VSS
B12	3200	4800	ddr[44]
B13	4000	4800	ddr[36]
B14	4800	4800	VSS
B15	5600	4800	VDD
C1	−5600	4000	SIG_B[25]
C2	−4800	4000	SIG_B[31]
C3	−4000	4000	SIG_B[26]
C4	−3200	4000	SIG_B[1]
C5	−2400	4000	VDD
C6	−1600	4000	SIG_A[29]
C7	−800	4000	SIG_A[33]
C8	0	4000	SIG_A[34]
C9	800	4000	SIG_A[37]
C10	1600	4000	VSS
C11	2400	4000	ddr[48]
C12	3200	4000	ddr[40]
C13	4000	4000	ddr[32]
C14	4800	4000	ddr[23]
C15	5600	4000	ddr[12]
D1	−5600	3200	VSS
D2	−4800	3200	SIG_B[32]
D3	−4000	3200	VDD
D4	−3200	3200	SIG_B[17]
D5	−2400	3200	VSS
D6	−1600	3200	SIG_A[12]
D7	−800	3200	SIG_A[14]
D8	0	3200	VSS
D9	800	3200	SIG_A[25]
D10	1600	3200	VDD
D11	2400	3200	VDD
D12	3200	3200	ddr[31]
D13	4000	3200	ddr[28]

(continued)

Table 6.6 (continued)

Pin number	x-coor	y-coor	Net name
D14	4800	3200	VSS
D15	5600	3200	ddr[13]
E1	−5600	2400	SIG_B[11]
E2	−4800	2400	SIG_B[2]
E3	−4000	2400	SIG_B[33]
E4	−3200	2400	SIG_B[18]
E5	−2400	2400	VDD
E6	−1600	2400	SIG_A[8]
E7	−800	2400	SIG_A[9]
E8	0	2400	SIG_A[10]
E9	800	2400	SIG_A[11]
E10	1600	2400	VSS
E11	2400	2400	ddr[35]
E12	3200	2400	ddr[27]
E13	4000	2400	ddr[24]
E14	4800	2400	ddr[2]
E15	5600	2400	ddr[3]
F1	−5600	1600	VDD
F2	−4800	1600	SIG_B[3]
F3	−4000	1600	VSS
F4	−3200	1600	SIG_B[27]
F5	−2400	1600	VDD
F6	−1600	1600	VSS
F7	−800	1600	VDD
F8	0	1600	VSS
F9	800	1600	VDD
F10	1600	1600	VSS
F11	2400	1600	VDD
F12	3200	1600	ddr[1]
F13	4000	1600	ddr[14]
F14	4800	1600	VDD
F15	5600	1600	ddr[4]
G1	−5600	800	SIG_B[12]
G2	−4800	800	SIG_B[34]
G3	−4000	800	SIG_B[19]
G4	−3200	800	SIG_B[4]
G5	−2400	800	VSS
G6	−1600	800	VDD
G7	−800	800	VSS
G8	0	800	VDD
G9	800	800	VSS

(continued)

Table 6.6 (continued)

Pin number	x-coor	y-coor	Net name
G10	1600	800	VDD
G11	2400	800	VSS
G12	3200	800	ddr[15]
G13	4000	800	VSS
G14	4800	800	ddr[16]
G15	5600	800	ddr[5]
H1	−5600	0	VSS
H2	−4800	0	SIG_B[35]
H3	−4000	0	VDD
H4	−3200	0	SIG_B[20]
H5	−2400	0	VDD
H6	−1600	0	VSS
H7	−800	0	VDD
H8	0	0	VSS
H9	800	0	VDD
H10	1600	0	VSS
H11	2400	0	VDD
H12	3200	0	ddr[6]
H13	4000	0	ddr[17]
H14	4800	0	ddr[18]
H15	5600	0	ddr[7]
J1	−5600	−800	SIG_B[6]
J2	−4800	−800	SIG_B[28]
J3	−4000	−800	SIG_B[13]
J4	−3200	−800	SIG_B[5]
J5	−2400	−800	VSS
J6	−1600	−800	VDD
J7	−800	−800	VSS
J8	0	−800	VDD
J9	800	−800	VSS
J10	1600	−800	VDD
J11	2400	−800	VSS
J12	3200	−800	ddr[9]
J13	4000	−800	ddr[19]
J14	4800	−800	VSS
J15	5600	−800	ddr[8]
K1	−5600	−1600	VDD
K2	−4800	−1600	SIG_B[36]
K3	−4000	−1600	VSS
K4	−3200	−1600	SIG_B[21]
K5	−2400	−1600	VDD

(continued)

Table 6.6 (continued)

Pin number	x-coor	y-coor	Net name
K6	−1600	−1600	VSS
K7	−800	−1600	VDD
K8	0	−1600	VSS
K9	800	−1600	VDD
K10	1600	−1600	VSS
K11	2400	−1600	VDD
K12	3200	−1600	ddr[25]
K13	4000	−1600	VDD
K14	4800	−1600	ddr[10]
K15	5600	−1600	ddr[20]
L1	−5600	−2400	SIG_B[8]
L2	−4800	−2400	SIG_B[14]
L3	−4000	−2400	SIG_B[22]
L4	−3200	−2400	SIG_B[7]
L5	−2400	−2400	VDD
L6	−1600	−2400	SIG_C[17]
L7	−800	−2400	SIG_C[16]
L8	0	−2400	SIG_C[15]
L9	800	−2400	SIG_C[14]
L10	1600	−2400	VSS
L11	2400	−2400	ddr[38]
L12	3200	−2400	ddr[33]
L13	4000	−2400	ddr[22]
L14	4800	−2400	VDD
L15	5600	−2400	ddr[21]
M1	−5600	−3200	VSS
M2	−4800	−3200	SIG_B[23]
M3	−4000	−3200	VDD
M4	−3200	−3200	SIG_B[37]
M5	−2400	−3200	VSS
M6	−1600	−3200	SIG_C[2]
M7	−800	−3200	VDD
M8	0	−3200	SIG_C[1]
M9	800	−3200	SIG_C[0]
M10	1600	−3200	VDD
M11	2400	−3200	VSS
M12	3200	−3200	ddr[37]
M13	4000	−3200	VSS
M14	4800	−3200	ddr[29]
M15	5600	−3200	ddr[11]
N1	−5600	−4000	SIG_B[39]

(continued)

Table 6.6 (continued)

Pin number	x-coor	y-coor	Net name
N2	−4800	−4000	SIG_B[24]
N3	−4000	−4000	SIG_B[9]
N4	−3200	−4000	SIG_B[29]
N5	−2400	−4000	VDD
N6	−1600	−4000	SIG_C[6]
N7	−800	−4000	SIG_C[5]
N8	0	−4000	SIG_C[4]
N9	800	−4000	SIG_C[3]
N10	1600	−4000	VSS
N11	2400	−4000	ddr[49]
N12	3200	−4000	ddr[42]
N13	4000	−4000	ddr[34]
N14	4800	−4000	ddr[30]
N15	5600	−4000	ddr[26]
P1	−5600	−4800	VDD
P2	−4800	−4800	VSS
P3	−4000	−4800	VSS
P4	−3200	−4800	SIG_B[38]
P5	−2400	−4800	VSS
P6	−1600	−4800	SIG_C[9]
P7	−800	−4800	SIG_C[8]
P8	0	−4800	VSS
P9	800	−4800	SIG_C[7]
P10	1600	−4800	VDD
P11	2400	−4800	VDD
P12	3200	−4800	ddr[46]
P13	4000	−4800	ddr[41]
P14	4800	−4800	VSS
P15	5600	−4800	VDD
R1	−5600	−5600	VSS
R2	−4800	−5600	VDD
R3	−4000	−5600	SIG_B[30]
R4	−3200	−5600	SIG_B[15]
R5	−2400	−5600	VDD
R6	−1600	−5600	SIG_C[13]
R7	−800	−5600	SIG_C[12]
R8	0	−5600	SIG_C[11]
R9	800	−5600	SIG_C[10]
R10	1600	−5600	VSS
R11	2400	−5600	ddr[50]
R12	3200	−5600	VSS

(continued)

Table 6.6 (continued)

Pin number	x-coor	y-coor	Net name
R13	4000	−5600	ddr[45]
R14	4800	−5600	VDD
R15	5600	−5600	VSS

6.3 Device Floorplanning and Co-design

Floorplanning is a process in the die design physical implementation flow where circuit blocks are arranged in the die. One of the major goals during the floor-planning process is to minimize the interconnect delay among the blocks and more importantly minimize the die size. In addition, clock tree and power distribution as well as I/O distribution are planned during floorplanning. From an ideal design perspective, package designers should collaborate with the device physical designers during the floorplanning process, in particular during device I/O distribution and placement. We refer to such mutual collaboration among the device designers and package designers during the floorplanning stage as silicon-package co-design.

The goal of silicon-package co-design process is simultaneous optimization of the device and package in terms of I/O placement to minimize total costs and to improve performance. Note that in many cases, package costs more than the device itself; thus, anything that minimizes the costs of the package is of great value. Package costs, in particular substrate, as discussed in previous chapters are governed by the substrate size, number of metal layers, substrate feature size such as via size, trace width/spacing, and selected materials. Die-package co-design methodology is aimed at reducing the package and die size, reducing routing congestion, and reducing the substrate layers by optimizing the I/O cell placement, device bump pattern, and package BGA ball pattern in the context of the overall system.

During the early stages of the co-design process, package engineer is expected to be able to create a virtual die, including the device I/O buffer cell libraries, bump pattern, substrate BGA ball pattern and perform pathfinding to propose an end-to-end optimal system connectivity.

In order to work closely with the die design team, package design team must have at least a high-level understanding of the die top-level physical layout and terminologies.

One of the building blocks of a device is the I/O buffer cell. An I/O buffer cell as depicted in Fig. 6.14 consists of I/O buffer area and I/O pad area. An I/O buffer cell can be of type signal, power, or ground net. I/O buffers can be placed on the periphery of the die, known as the I/O pad ring, Fig. 6.15. The I/O pads are routed

Fig. 6.14 An I/O buffer cell consists of I/O buffer and I/O pad

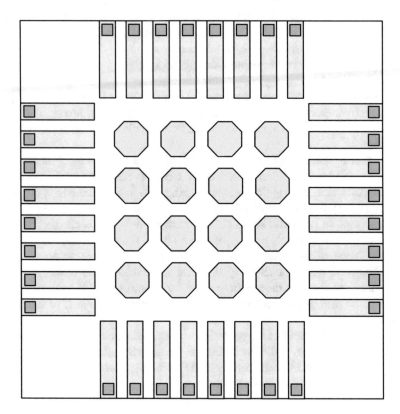

Fig. 6.15 A typical simple representation of I/O buffer placement on the die periphery (I/O pad ring) and bump pads at the center

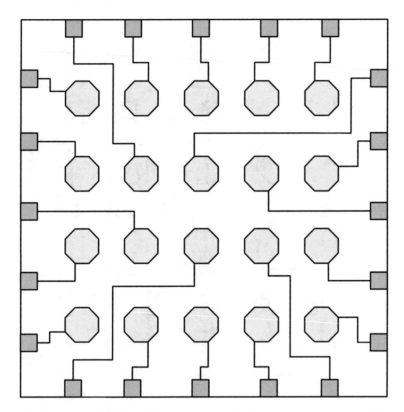

Fig. 6.16 I/O pads on the die periphery are connected to bump pads at the center using on die RDL layers

to bump pads using die top metal layers, also known as redistribution layer (RDL layer) Fig. 6.16. Likewise, I/O pads maybe placed on the entire die surface area, resulting in shorter RDL length and improved signal and power integrity (Fig. 6.17).

Similar to Example-1 where we optimized the die bump pattern with respect to BGA ball pattern, we also need to optimize the I/O pad placement with respect to bump pad or vice versa to avoid wire crossing in RDL layers. I/O pad placement and die size are inter-related; an alternative I/O pad placement (Fig. 6.18) may result in smaller die size; although smaller die size in general may sound attractive, smaller die size may result in adding more layers to the package substrate producing a more expensive end product compared to a slightly larger die. Thus, I/O pad placement and its effect to RDL routing and package substrate routing must be investigated in depth at the early stages of product design.

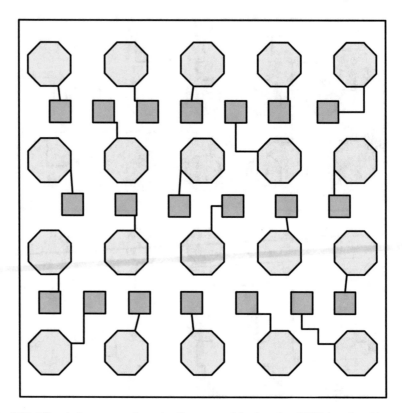

Fig. 6.17 I/O pad placement on the entire die area, resulting in reduced RDL length and improved signal/power integrity

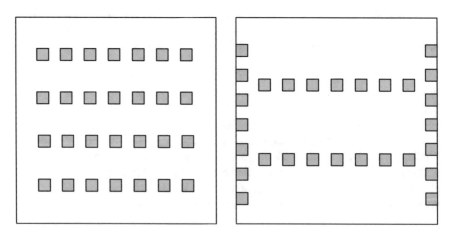

Fig. 6.18 Various examples of I/O pad placement

Pathfinding and design optimization become increasingly important and essential part of any heterogeneous integration where multiple devices of various sizes, process nodes, interfaces, and bump patterns must be integrated on a fine pitch silicon interposer.

Co-design and Pathfinding become increasingly important and are an integral part of any heterogeneous integration where multiple devices of various sizes, process nodes, and bump patterns are integrated on a fine pitch silicon interposer and organic substrate. Next, we will explore in more detail some methodologies for pathfinding and optimization of various 2.5D/3D scenarios.

6.4   Pathfinding Methodology for Optimal Design and Integration of 2.5D/3D Interconnects [1, 2]

One of the key challenges in designing a low-cost and high-performance 2.5D/3D package is the system-level ultra-dense connectivity exploration and pathfinding. Planning an I/O pad ring structure and defining I/O buffer cell placement for multiple logic partitions at the early stages of 2.5D/3D product development are not trivial. Moreover, end-to-end optimization of inter-partitions I/O cell placement, Cu pillar bump matrix, and package BGA pattern interfaced with numerous components on the PCB with fixed ball patterns is a daunting task.

We will explore a pathfinding methodology for the design and optimization of 2.5D/3D interconnects. We will explore this methodology in a 2.5D/3D design where inter-partitions I/O buffer cells placement, Cu pillar bump matrix, and package BGA pattern are optimized in a hierarchical fashion with respect to the wide I/O memory's fixed bump pattern interfaced with the PCB level components. We will also look at the cross-domain flexibility and robustness of this methodology by performing an end-to-end pathfinding on ultra-dense 2.5D/3D silicon interposer and single monolithic device interfaced with fixed components on a PCB.

A 2.5D/3D pathfinding and integration methodology must be capable of performing an early feasibility and evaluation of ultra-dense device connectivity, routability, risks, performance, and cost in the context of multiple system configuration (Fig. 6.19).

We demonstrate an end-to-end pathfinding and optimization methodology for the design and optimization of a 2.5D/3D system interconnect with capabilities for simultaneous exploration of various package types, various stacking technologies, various assembly technologies, and various PCB technologies considering performance criteria such as signal/power integrity, timing, temperature.

We demonstrate this methodology by integrating wide I/O memory and multi-partition logics on a 2.5D/3D package interfaced with a monolithic SOC flip chip die realized on a multi-layer organic build-up substrate. The flip chip SOC package is subsequently interfaced with a rigid silicon interposer package, a DDR4 package, a connector, and a wirebonded BGA laminate package on the PCB (Fig. 6.20). Finally, we will perform an end-to-end pathfinding and optimization on

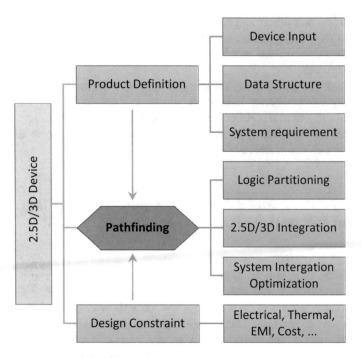

Fig. 6.19 Pathfinding in 2.5D/3D system design

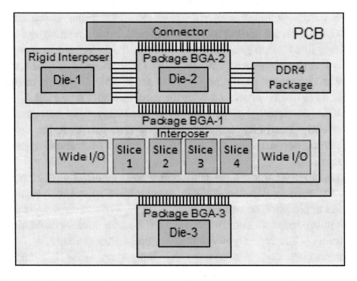

Fig. 6.20 A 2.5D/3D package interfaced with a flip chip package, a rigid interposer package, a DDR4 Package, a connector, and a wirebonded package

the overall system. In particular, the robustness of our methodology strives for early evaluation of routability, risks, cost, performance, and product development.

6.4.1   Methodology

Normally, multi-layer build-up substrate contributes to more than 40% of the total package cost; thus, a poorly designed substrate can easily exceed the cost of the die it contains; this problem is amplified considering a 2.5D/3D silicon interposer-based package. In system design, often one or more packages come with a fixed ball pattern, and this is normally due to customer requirement/specification, system constraint, legacy product, or an off-the-shelf component.

As noted previously, integrating multiple components where one or more component has a fixed bump/ball pattern creates a need for pathfinding and co-design methodology. The ultimate goal is to design a 2.5D/3D device so that its bump pattern is optimized with respect to the system components with fixed ball patterns (off-the-shelf packages); more precisely, the 2.5D/3D partitions I/O buffer cell placement must be optimized in the context of system components with fixed ball patterns. This means that components with fixed ball pattern drive the system connectivity (Fig. 6.21). In this demonstration, components with fixed ball pattern/netlist are wide I/O memory, DDR4, and connector. We will show how I/O buffer cells placement within the device can be optimized to match the components with fixed pattern mounted on both interposer and PCB.

Fig. 6.21 2.5D/3D pathfinding flow. Components with fixed ball pattern drive the system connectivity

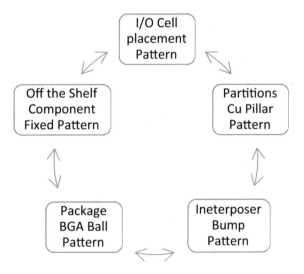

6.4.2   Wide I/O Memory

Wide I/O memory is a low-power device, targeted mainly for mobile/tablet market. It comes with fixed Cu pillar pattern; pads are located at the center of the die and defined by JEDEC standards (JESD229). Wide I/O-1 has large bus interface (512 bits), four independent asynchronous channels, pillar pitch of 50 µm horizontal and 40 µm vertical, with total of ~ 1200 pillars.

6.4.3   Pathfinding and Optimization

Pathfinding and optimization sequence for this study are illustrated in Fig. 6.22. Considering system development is driven from the wide I/O memory perspective; we begin the journey with placement of wide I/O memory dies and logic partitions as described in Fig. 6.22. Slices 1 and 4 are optimized with respect to wide I/O memory die (Fig. 6.23) followed by optimizing slices 2 and 3 with respect to slices 1 and 4. Subsequently, BGA-1 package and interposer are optimized with respect to logic slices (Figs. 6.24 and 6.25).

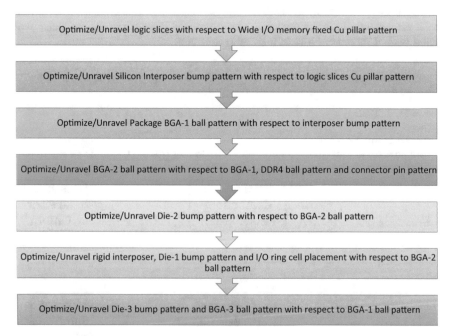

Fig. 6.22 Pathfinding and optimization sequence for integrating components depicted in Fig. 6.20

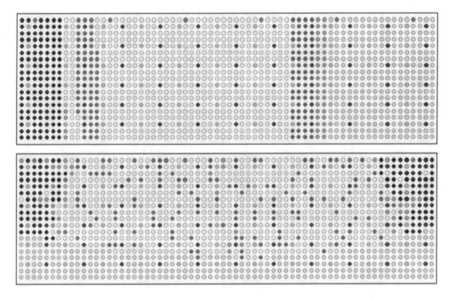

Fig. 6.23 Logic partition interfaced to wide IO memory. (Top) before optimization, (Bottom) after optimization

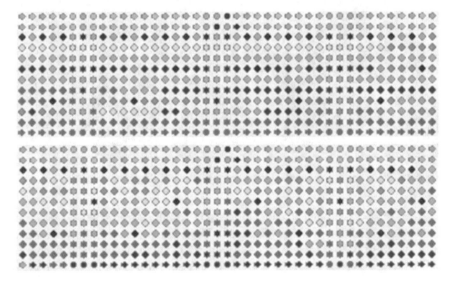

Fig. 6.24 Portion of silicon interposer depicting, (Top) before optimization, (Bottom) after optimization

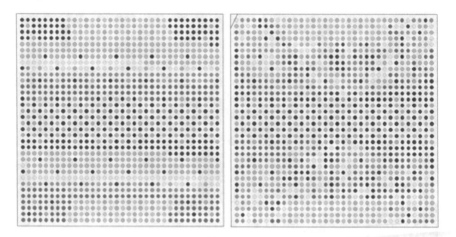

Fig. 6.25 BGA-1 package ball-out. (Left) before optimization, (Right) after optimization

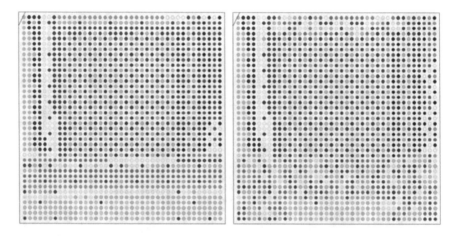

Fig. 6.26 BGA-2 package ball-out. (Left) before optimization, (Right) after optimization

Next, BGA-2 package ball pattern is optimized with respect to BGA-1 package, connector and DDR-4 package (Fig. 6.26). Die-2 bump pattern is then optimized with respect to BGA-2 package ball-out (Fig. 6.27). Subsequently, rigid interposer ball pattern is optimized with respect to BGA-2 package ball-out (Fig. 6.28)

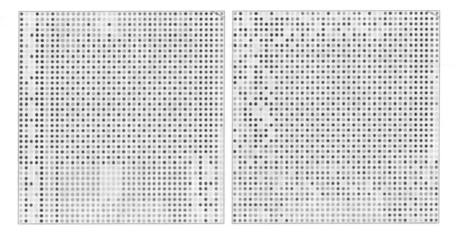

Fig. 6.27 Die-2 bump assignment. (Left) before optimization, (Right) after optimization

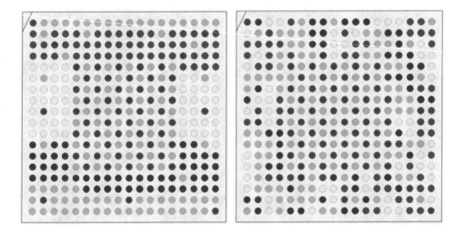

Fig. 6.28 Rigid Interposer package ball-out. (Left) before optimization, (Right) after optimization

followed by optimizing the Die-1 bump pattern with respect to rigid interposer ball-out (Fig. 6.29). Next, we will optimize the I/O buffer cell placement with respect to Die-1 bump pattern (Fig. 6.30). At this point, all components situated above the BAG-1 package depicted in Fig. 6.20 should be fully optimized with respect to BGA-1 package (Fig. 6.31).

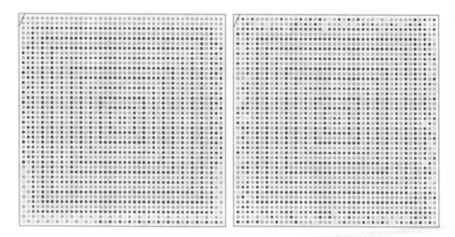

Fig. 6.29 Die-1 bump pattern optimization. (Left) before optimization, (Right) after optimization

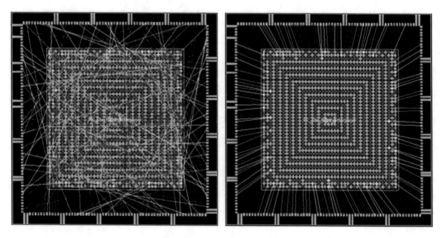

Fig. 6.30 Optimizing Die-1 I/O buffer cell placement with respect to bump pads. (Left) before optimization, (Right) after optimization

Finally, BGA-3 package ball pattern is optimized with respect to BGA-1 package ball-out (Fig. 6.32) followed by optimizing the Die-3 bonding pads with respect to BGA-3 package ball pattern (Figs. 6.33 and 6.34). Figure 6.35 illustrates the optimized flightline connectivity between Die-3, BGA-3, and BGA-1 package.

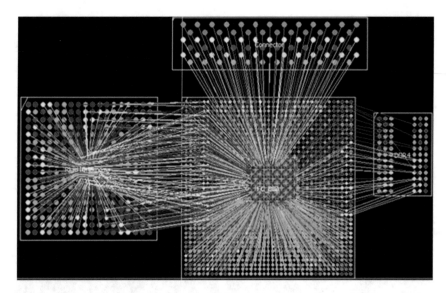

Fig. 6.31 Rigid silicon interposer, Die-1, Die-1 I/O buffer cells, connector, DDR4 package, Die-2, and BGA-2 package fully optimized with respect to BGA-1 package

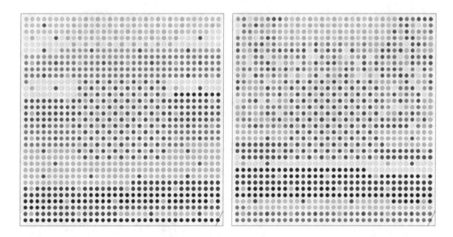

Fig. 6.32 BGA-3 package ball pattern optimized with respect to BGA-1 package. (Left) before optimization, (Right) after optimization

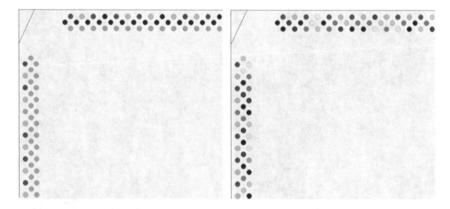

Fig. 6.33 Die-3 bonding pads are optimized. (Left) before optimization, (Right) after optimization

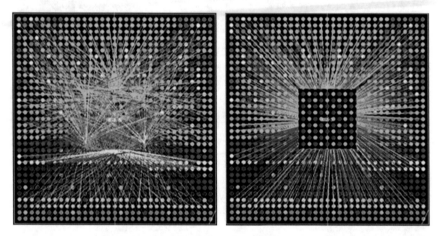

Fig. 6.34 Die-3 optimized with respect to BGA-3 package. (Left) before optimization, (Right) after optimization

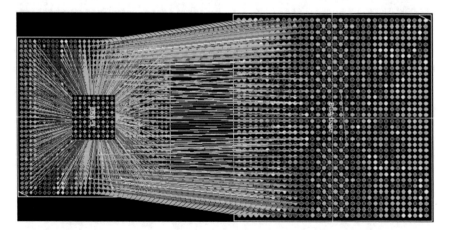

Fig. 6.35 Die-3 and BGA-3 package are optimized with respect to BGA-1 package

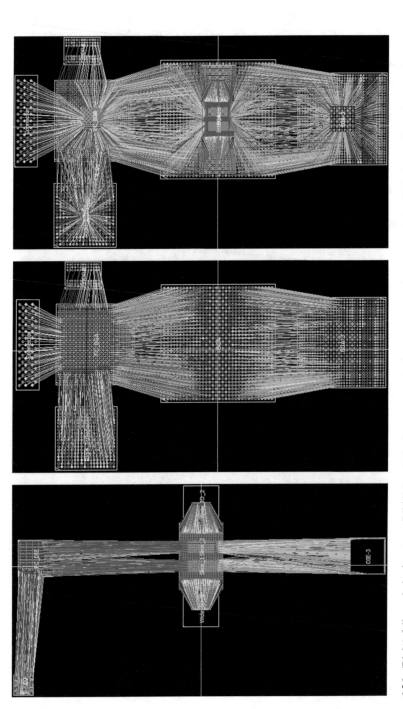

Fig. 6.36 (Right) fully optimized system, (Middle) optimized components ball pattern across the system, (Left) all devices optimized across the system

At this point, the system should be fully optimized as illustrated in Fig. 6.36. A full system optimization in this case includes optimization across the devices and components. Figure 6.37 shows the fully optimized high-speed Ser/Des differential pairs across the devices and components. Shielding of these pairs was enforced by applying shielding constraints during placement.

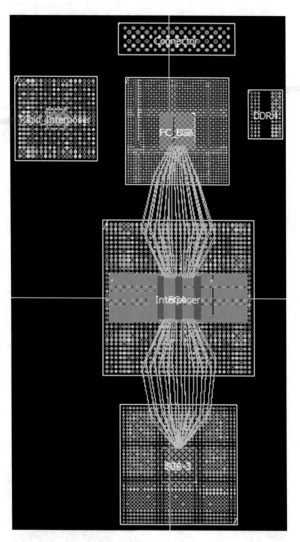

Fig. 6.37 High-speed Ser/Des differential pairs are fully optimized across devices and components

6.5   Pathfinding and Design Optimization of 2.5D/3D Devices in the Context of Multiple PCBs [1, 2]

2.5D/3D products can be very costly if not designed optimally. Elements of cost in 2.5D/3D integrations are process node technology, number of metal layers, packaging, reliability, yield, and IP reuse. Cost and performance are directly tied to pathfinding and physical design optimization. The objective of pathfinding and optimization is to produce a reliable, low-cost and miniaturized system with high performance (Fig. 6.38).

Slicing a 2D monolithic mixed-signal system-on-chip (SOC) die into multiple logic partitions and subsequent heterogeneous integration results in an ultra-dense off-chip connectivity. Defining, planning, and managing a 2.5D/3D device with ultra-dense connectivity at the early stages of product development is a new challenge that the industry is facing today.

One of the major requirements to perform an early 2.5D/3D product feasibility analysis and ultimately successful cost-effective product design is to have a database capable of unifying the chip I/O buffer cells placement, interposers, packages, and PCBs. With such unified database, one can explore routability, risks, cost and effective early stage evaluation and optimal design of 2.5D/3D devices with ultra-dense connectivity (Fig. 6.39).

Fig. 6.38 Objective of pathfinding and optimization in system design

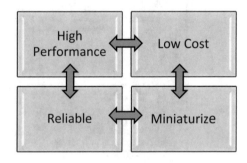

Fig. 6.39 A unified database for pathfinding and design optimization of 2.5D/3D systems

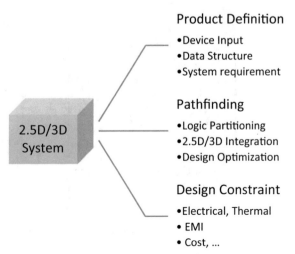

With the industry moving toward 2.5D/3D integration, it is no longer economical to simply lay out a package substrate that meets certain ball-out and electrical requirement. 2.5D/3D devices are super sensitive with respect to cost and performance.

The objective is to demonstrate pathfinding and design optimization methodology for optimal integration of 2.5D/3D devices in the context of multiple systems. We demonstrate a methodology for integrating a 2.5D/3D device in the context of multiple PCB systems.

6.5.1 Multiple Integration Scenarios

Considering the cost versus performance requirement, it is becoming increasingly important to be able to investigate and perform pathfinding and design optimization of a device in the context of multiple PCBs (Fig. 6.40).

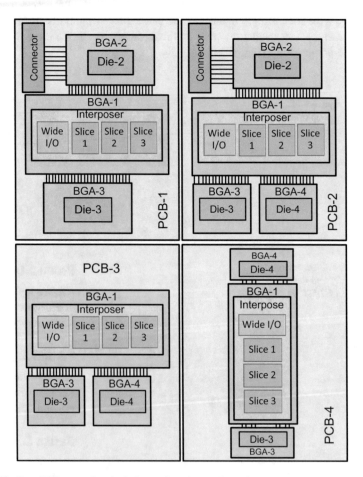

Fig. 6.40 Four PCB scenarios depicting various integration schemes

6.5.1.1 PCB-1 Scenario

In this scenario, vendor supplying BGA-1 package is deriving the total system design. Components with fixed netlist are wide I/O memory and connector. Logic slices, interposer, BGA-1, BGA-2, BGA-3, Die-2, Die-3 will be optimized with respect to components with fixed netlist.

6.5.1.2 PCB-2 Scenario

In this scenario, vendor supplying BGA-1 owns the total system design. Components with fixed netlist are connector, wide I/O memory. Interposer, BGA-1, BGA-3, and BGA-4 substrates will be optimized with respect to components with fixed netlist. This demonstrates the concept of two package solution compared to one package solution initially demonstrated in PCB-1 scenario.

6.5.1.3 PCB-3 Scenario

In this scenario, vendor supplying BGA-3, and BGA-4, is deriving the total system design. BGA-1, interposer and logic slices, BGA-3, and Die-4 will be optimized with respect to wide I/O, Die-3, and BGA-4. Components with fixed netlist are wide I/O memory, Die-3, BGA-4.

6.5.1.4 PCB-4 Scenario

In this scenario, we demonstrate the effect of package placement on pathfinding and optimal pin assignment. Vendor in charge of designing the system is proposing placement of BGA-3 and BGA-4 on the corners of BGA-1 package. Considering such placement scenario BGA-1, interposer and logic slices will be optimized with respect to proposed placement position. Components with fixed netlist are wide I/O, BGA-3, and BGA-4.

6.5.2 Methodology

If not designed optimally, organic build-up substrates are major contributor to overall package cost. In flip chip packaging, normally, more than 40% of the total package cost is attributed to build-up substrate. This phenomenon can be amplified by adding an interposer to the total equation. Thus, an optimal pathfinding methodology that produces a substrate with fewer metal layers and higher performance is of great value.

Designing a system involves integrating multiple components from various suppliers on a PCB. Off-the-shelf components are inherently fixed with respect to netlist, package size, and solder ball pitch. Thus, a new IC design begins with pathfinding and optimizing the I/O buffer cell placement with respect to the system components with fixed netlist and footprint (Fig. 6.41).

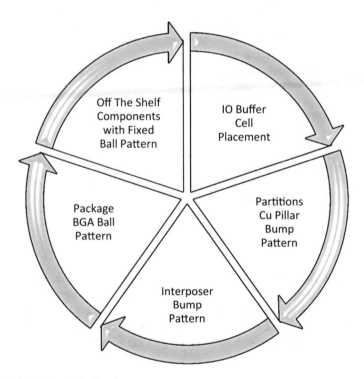

Fig. 6.41 2.5D/3D pathfinding flow

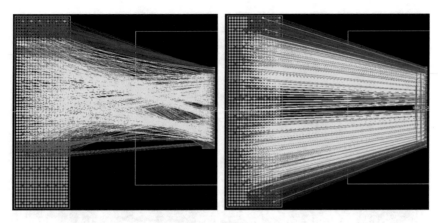

Fig. 6.42 Wide I/O to logic slice, (Left) before optimization, (Right) after optimization

Considering PCB-1 scenario, we begin the optimization process by:

(A) optimize slices 1 bump pattern with respect to wide I/O memory bump pattern (Fig. 6.42),
(B) optimize slice 1, slice 2, and slice3 bump patterns; these are the parallel bus interfaces (Fig. 6.43),
(C) optimize interposer bump pattern with respect to slices-2 and slice-3 bump pattern (Fig. 6.44),
(D) optimize BGA-1 ball pattern with respect to interposer bump pattern (Fig. 6.45),
(E) optimize BGA-2 ball pattern with respect to BGA-1 ball pattern and connector pin pattern,
(F) optimize Die-2 bump pattern with respect to BGA-2 ball pattern,
(G) optimize BGA-3 ball pattern with respect to BGA-1 ball pattern,
(H) optimize Die-3 I/O buffer cells placement with respect to BGA-3 ball pattern.

Complete PCB-1 connectivity before and after optimization is illustrated in Fig. 6.46.

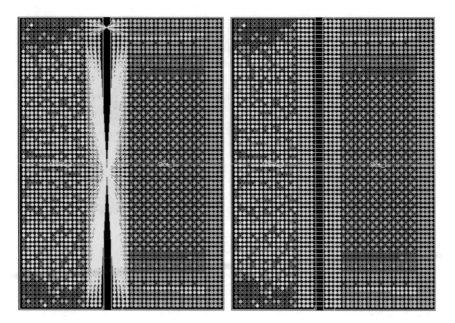

Fig. 6.43 Slice 1 to slice 2, (Left) before optimization, (Right) after optimization

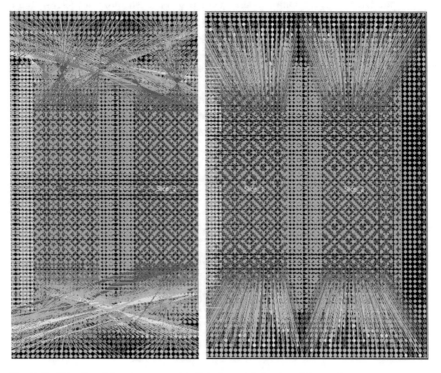

Fig. 6.44 Slice 2 and 3 to interposer, (Left) before optimization, (Right) after optimization

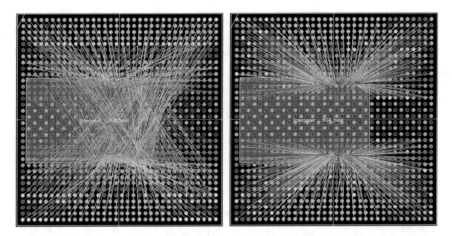

Fig. 6.45 Interposer to BGA-1, (Left) before optimization, (Right) after optimization

Steps to carry out pathfinding and optimization for PCB-2, PCB-3, and PCB-4 scenarios are similar to PCB-1 scenario. Figure 6.47 shows the concept of two package solution; Interposer, BGA-1, BGA-3, and BGA-4 substrates are optimized with respect to components with fixed netlist (PCB-2 scenario). Figures 6.48 and 6.49 demonstrate the effect of package placement/orientation on interposer and logic partitions bump pattern as specified in PCB-3 and PCB-4 scenarios. Figure 6.50 shows the optimized device connectivity across the system.

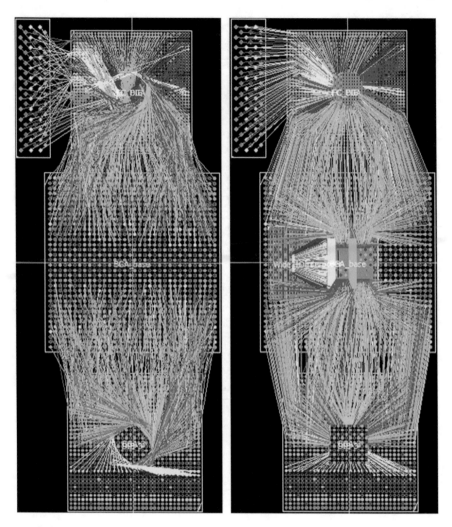

Fig. 6.46 Complete optimization of PCB-1 depicted in Fig. 6.40, (Left) before optimization, (Right) after optimization

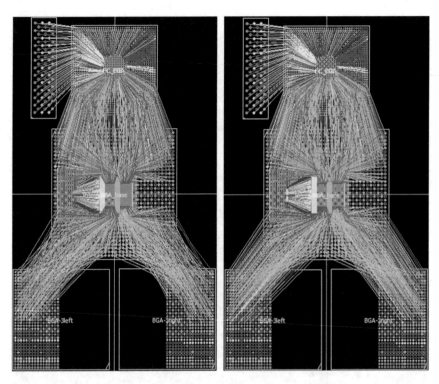

Fig. 6.47 Complete optimization of PCB-2 depicted in Fig. 6.40, (Left) before optimization, (Right) after optimization

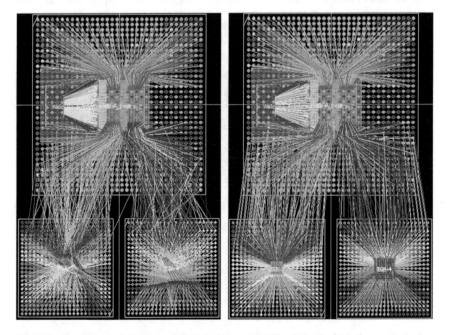

Fig. 6.48 Complete optimization of PCB-3 depicted in Fig. 6.40, (Left) before optimization, (Right) after optimization

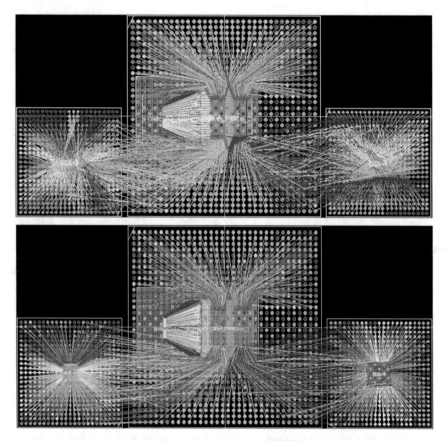

Fig. 6.49 Complete optimization of PCB-4 depicted in Fig. 6.40, (Top) before optimization, (Bottom) after optimization

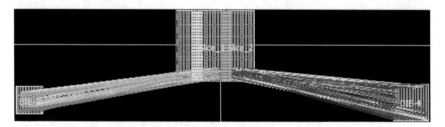

Fig. 6.50 Optimization of PCB-4 depicted in Fig. 6.40, demonstrating optimization of Die-3 and Die-4 to logic slices

References

1. Yazdani F, Park J (2014) Pathfinding and design optimization of 2.5D/3D devices in the context of multiple PCBs. In: IMAPS 10th international conference on device packaging, Fountain Hills, AZ, USA, pp 294–297, 10–13 Mar 2014
2. Yazdani F, Park J (2014) Pathfinding methodology for optimal design and integration of 2.5D/3D interconnects. In: Proceedings of the 64th IEEE electronic components and technology conference, Orlando, Florida, 26–30 May 2014

Index

© Springer International Publishing AG 2018
F. Yazdani, *Foundations of Heterogeneous Integration: An Industry-Based,*
2.5D/3D Pathfinding and Co-Design Approach,
https://doi.org/10.1007/978-3-319-75769-8

Printed in the United States
By Bookmasters